"Erin's easy-to-follow instructions on milling your own flour will have even the most inexperienced cook running for the kitchen."

—Ashley McLaughlin, author of *Baked Doughnuts for Everyone*

"In this comprehensive and inspired book, you'll find an impressive range of sweet and savory recipes, with concise and accessible information on milling your own grains. This is a beautiful cookbook that will rise to the top of my stack, one that I'll return to again and again."

—Megan Gordon, author of *Whole-Grain Mornings*

THE HOMEMADE FLOUR COOKBOOK

The Home Cook's
Guide to Milling Nutritious
Flours and Creating
Delicious Recipes with
Every Grain, Legume, Nut,
and Seed from A-Z

ERIN ALDERSON

Creator and founder of the blog
Naturally Ella

Fair Winds Press
100 Cummings Center, Suite 406L
Beverly, MA 01915

fairwindspress.com • quarryspoon.com

First published in the USA in 2014 by
Fair Winds Press, a member of
Quarto Publishing Group USA Inc.
100 Cummings Center
Suite 406-L
Beverly, MA 01915-6101
www.fairwindspress.com
Visit www.quarryspoon.com and help us celebrate food and culture one spoonful at a time!

18 17 16 15 14   2 3 4 5

ISBN: 978-1-59233-600-5

Digital edition published in 2014
eISBN: 978-1-62788-017-6

Library of Congress Cataloging-in-Publication Data available

Cover and book design by Kathie Alexander
Book layout by Sporto
Photography by Erin Alderson

Printed and bound in China

*The information in this book is for educational purposes only. It is not intended to replace the advice of a
physician or medical practitioner. Please see your health care provider before beginning any new health program.*

To my love, Mike.
Many great things are possible
with you by my side.

CONTENTS

INTRODUCTION

My Unprocessed Kitchen

My childhood was one of a typical American middle-class family. We were always on the move with activities and concerts. Sit-down meals were few and far between. I ate my fair share of fast food and didn't think twice about where my food was coming from; I just knew I liked to eat. I felt only the faintest connection to my food.

This all began to change when I was nineteen, between my freshman and sophomore years of college. Like most first-year college students, I ate my way through that year and was the heaviest and unhealthiest I had been in my life. That summer, however, my father suffered a major heart attack at the age of 45 and endured a quadruple bypass surgery. Watching my father go through that experience started me down a road of questioning my eating choices. With a major push from my grandmother, I started focusing on health for the first time in my life.

The first step, of many baby steps, was to weed out fast food and start focusing on the quality of the food I was consuming. I threw myself into learning about nutrition, and eventually these baby steps toward health helped me discover my love of cooking. I began experimenting more in the kitchen, veering away from recipes and using the ingredients I had on hand. I visited the farmers' markets and joined a CSA (Community Supported Agriculture), both of which played a large role in my quest for unprocessed foods. The more I learned about food and cooking, the more I moved away from store-bought processed foods.

That journey has led me here. I try to make everything from scratch, using the freshest ingredients. So many products in supermarkets are overly processed and stripped of their nutrients. After making all those steps toward fresh, minimally processed foods, the prepackaged products became less appealing. My refrigerator is full of fruits, vegetables, and locally sourced dairy products, while my cupboards contain my own miniature bulk-bin section of foods.

My exploration of cooking led to the start of my blog, Naturally Ella. I wanted to share recipes that were based around minimally processed food. The more I shared my journey, the more I started learning. I looked into flour, especially wheat flour, and found not only that it was often cheaper to make at home but also that the flavor was better. I continued my exploration with more items from my pantry, especially with gluten-free and nut flours. This book is a culmination of all I've learned about grinding flours at home.

Although my journey is important in the context of why I wrote this book, I believe everyone's food journey is unique and can be very personal. My hope is that this book will provide helpful information and guidance for creating your own minimally processed kitchen, along with inspiration to look at ingredients you already have in a new way.

CHAPTER 1

Milling at Home

I started milling flours at home for two reasons: I wanted to have fresh, unrefined flour, and I needed a use for all the grains, beans, and nuts I loved buying in bulk. Often, packaged flours at the grocery store are stripped of a many, if not all, of their nutrients, so milling flours at home is a great way to add more nutrition to your diet.

However, often the flavors and textures of home-milled flours are not quite what you might expect. For example, some flours soak up more moisture, which can affect the outcome of a recipe. Learning to work with freshly milled flours is a game of experimentation and adapting to new tastes and textures. I've often heard that people give up eating whole wheat bread because it does not taste like white bread. But it's not supposed to! Instead, wheat should be celebrated for its own flavor.

It should be no surprise, then, that all flours are not created the same. Certain wheat varieties, for example, are prized for their amazing protein content, which creates rise and lift in breads; so substituting other flours for high-protein flours can sometimes lead to disappointment. That's why I use a mixture of flour and starch in gluten-free baking, which can help achieve the characteristics I'm looking for in the finished product. That said, I also like to experiment with textures and flavors, so I embrace the grain for its unique characteristics, rather than forcing it to conform to a white flour standard. If you find that the taste or texture of a particular flour in this book isn't to your liking, try blending a couple flours together. Experimenting can be fun!

HOME MILLING EQUIPMENT

Before you start milling at home, you'll want to acquire a few key pieces of equipment. Some, such as an electric grain mill, may not be feasible purchases, but, luckily, some of the small appliances you already own may be up to the task of milling.

Electric Grain Mill: Without fail, every time I get into a conversation with someone about cooking at home, I launch into a diatribe about how wonderful it is to have an electric grain mill at home. Within minutes I can have freshly ground flour at my disposal that is more nutritious and usually a bit cheaper than buying flour from the store. I look at purchasing a grain mill as an investment. However, an electric grain mill won't grind nuts and seeds (see A, B, C).

Hand Mill: The non-electric, slightly cheaper, cousin of the electric grain mill can power through the same items as an electric mill and even more. Using this mill takes a bit more work and time but still provides you with nutritious, home-ground flour. The flour will be slightly coarser than flour from an electric mill but can be used interchangeably in recipes.

High-Powered Blender: These blenders are not your average blender but instead were created to tackle just about everything. Often these blenders can grind grains, legumes, nuts, and seeds into flour just as well as the grain mills can. These blenders are often a bit more expensive than electric grain mills but, of course, serve more than one purpose.

A

B

C

Food Processor/Blender: My food processor is one of the most important small appliances in my kitchen. It's versatile enough that I can whip up pie dough or vegetable burgers and then turn around and make flour. I use my food processor to make nut flours and oat flour from rolled oats (see A, B, and C). A regular blender could also tackle these jobs. When using the food processor, I always sift my flour after grinding (see D). This produces a lighter flour that works well in recipes.

Nut Mill: This inexpensive grinder does exactly what the name implies: It grinds nuts. As you read through the chapter on nuts (page 171), it will become apparent how easy it is to go from nut meal to nut butter in the blink of an eye. Using a nut mill can help prevent that from happening.

Coffee/Spice Grinder: My coffee grinder serves another purpose in my kitchen other than grinding coffee. I also use it to grind seeds, because they are too small for the food processor. It can grind small batches of grains as well.

Kitchen Scale: Weight is one of the most important factors when dealing with flours. As I quickly found out, not every measuring cup is created equal. I have four different sets of measuring cups, and only one is close to being accurate. For years I measured my flours by digging the scoop into the bag and patting down to level out the flour, which is completely wrong. I believe that many of my failed baking experiments in past years can be traced back to incorrect measuring. The best approach is to invest in a kitchen scale. I purchased a $30 scale 3 years ago that is still going strong. If you still plan on using measuring cups, the key is to spoon the flour into the cup, then level it off with a butter knife; do not pack it down. Between this method and the scale, your measurements should be spot on.

Odds and Ends: A good stand mixer or hand mixer is always a lovely addition to any kitchen. Many bread recipes in this book call for using the stand mixer; however, all can be made by hand with a large bowl, a solid wooden spoon, and your hands. I always have a pizza stone in my oven, and I would be lost without the pasta attachment for my stand mixer and an inexpensive ravioli cutter. I also keep a pastry mat handy to make kitchen cleanup easier and a silicone baking sheet liner to use in place of parchment paper.

ABOUT SUBSTITUTIONS

I think one of the most fun things to do in the kitchen is play. I rarely follow a recipe and am often switching out ingredients or adding extra flavors. Being able to do this comfortably is extremely useful with flours. Sometimes I'll change flours for taste or because I'm missing one that I may need.

Take, for example, the recipe that follows. It's a fairly basic pancake recipe that can be easily adapted to what you have on hand. Many times I'll mix different flours together for a multigrain pancake; 40 g each of spelt, rye, and oats make a lovely combination. For the most accurate substitutions, use weights, because some flours have different weights.

Substitutions can come in handy when converting a recipe to be gluten-free. For the majority of gluten-free baked recipes, I substitute two-thirds flours and one-third starch. The ratios change for yeasted products and piecrust because more starch is needed to keep a light consistency. For this particular recipe, I like 30 g oat flour, 30 g sorghum flour, 30 g millet flour, 15 g arrowroot, and 15 g cornstarch. The possible combinations of flours are many, but they're highly dependent on personal taste. Try a few different combinations to find the flours you like together! (Once you find a combination you like, try blending a big batch to use throughout the week.)

Throughout the book, there are helpful suggestions for individual flours and substitutions, because some flours act differently in the amount of liquid needed.

WHOLE WHEAT PANCAKES

Serve these pancakes warm with butter,
maple syrup, and fresh fruit.

1 cup (120 g) soft wheat flour

1 teaspoon baking powder

¼ teaspoon sea salt

1 tablespoon (12 g) butter, melted and slightly cooled

1 tablespoon (15 ml) maple syrup

2 large eggs

⅔ cup (160 ml) 2% milk

1 teaspoon vanilla extract

To keep pancakes warm before serving, preheat the oven to 200°F (100°C, or very low) with an ovenproof plate inside.

In a bowl, stir together the flour, baking powder, and salt. In a separate bowl, whisk together the melted butter, maple syrup, eggs, milk, and vanilla. Pour the wet ingredients into the dry ingredients and stir until just combined. Let sit for 5 minutes.

Preheat a lightly greased skillet or griddle over medium-low heat. Pour a scant ¼ cup batter onto the heated surface. Let cook for 2 to 3 minutes, until the batter begins to form bubbles. Flip the pancake and cook for another minute. Repeat with the remaining batter. Stack pancakes on the plate in the oven as they finish cooking.

PANTRY STAPLES

I keep regular stock of certain items in my house so that I can always have a meal on hand. All of these items are used throughout the book.

Fats and Oils: I like to keep this list fairly simple: butter, olive oil, walnut oil, almond oil, coconut oil, and occasionally sunflower oil. I keep a healthy stock of nut oils on hand to use in place of canola or safflower oil. I enjoy the slightly nutty hint they give to baked goods and salad dressings. I use butter for baked goods and smearing on pancakes and crepes, while I use olive oil for roasting and sautéing vegetables.

Sweeteners: In the past few years I've been pushing more toward using natural sweeteners such as honey, maple syrup, sorghum, and molasses. However, there are still a few recipes that benefit from organic cane sugar, brown sugar, and confectioners' sugar. I try to use them in moderation, substituting one of the less-refined natural sweeteners when I can.

Starches: Starches serve a dual purpose in my kitchen: as a thickening agent and as a partner to gluten-free flours. Although starches aren't always necessary, they do help to create a lighter, more traditional texture when using gluten-free flours. I use different mixes of arrowroot, tapioca, and cornstarch plus sweet brown rice flour (page 84).

Dairy: Dairy is a category that I am very picky about. I try my best to buy organic and local, and I use full-fat versions of yogurt, milk, and cheese when I can. Low-fat and nonfat versions lack the flavor of their full-fat counterparts. My husband and I have compromised on 2 percent milk, as he doesn't like the creaminess of whole milk. For recipes calling for 2 percent milk, whole milk can easily be used. Also, I use Greek yogurt instead of regular yogurt, primarily for its added creaminess, and as a great substitution for sour cream.

Legumes: While I cover these extensively in chapter 4 (page 135), I want to mention beans and weighing here. If you choose to use canned beans, note that all measured weights for cooked beans in the recipes were taken after the beans were rinsed and drained.

A NOTE ON STORAGE

As you read through this book, you'll see storage tips sprinkled throughout. However, I keep to a few general guidelines based on experience. For most items, it's hard to give a precise expiration date, such as "wheat berries will keep for exactly 1 year." When purchasing from bulk bins, it can be difficult to guess how long ago the item was originally packaged and how long it has been sitting in the bins.

I store all my grains and beans in glass jars with an airtight seal. Some say that intact grains and beans have a 4- to 6-month shelf life when stored at room temperature and a longer life when stored in the freezer. However, I've successfully kept some at room temperature for a year or two because I was able to avoid heat and moisture. If you are unsure, inspect the grains before using, checking for mold, bugs, or a spoiled, musty smell. Of course, the easiest way to avoid this issue is to buy in smaller quantities.

As for flours, once the grain, legume, or nut has been ground, the storage time drops. I keep all freshly ground flours in the refrigerator for up to 2 months, but occasionally less. These circumstances are noted throughout the book.

HOW TO MILL
AT HOME

Milling flour at home may seem a bit daunting, but the process goes rather quickly. Below are the general guidelines for grinding flour using an electric grain mill, food processor, and coffee grinder. I've noted specific grinding information with each item in the chapters.

Electric Grain Mill: Pick through the grains or beans for extraneous items, such as rocks or other debris. Pour the grains or beans into the grain mill according to the manufacturer's directions. Check and sift the flour for any unground pieces, especially with smaller grains such as amaranth and teff.

Food Processor: Add the oats, nuts, or other item to the food processor and pulse until they begin to break down into part flour, part meal. Pour through a sieve into a bowl and then return the non-flour part (the part in the sieve) back to the food processor and grind again. Repeat until whatever it is you're grinding is primarily in flour form. This process is particularly important when grinding nuts, to make flour rather than nut butter.

Coffee Grinder: Fill the coffee grinder roughly half full. Pulse the item until flour begins to form. Remove, sift through a sieve, and return the non-flour part back to the coffee grinder. Repeat until the item is mostly in flour form (see A, B, and C).

A

B

C

CHAPTER 2

Glorious Grains

In recent years, grains that contain gluten have gotten a slightly bad reputation because of the gluten-free diet craze and its bevy of associated products. But gluten, on its own, isn't a bad thing for those whose bodies can tolerate it. And there are some wonderful gluten-containing grains, quite a few of which now come with the distinction of being "ancient grains." These are grains that have stood the test of time, unchanged, for thousands of years. They are often praised for their high level of nutrients compared with more modern grains.

One of the most important facts to know about these glutenous grains is that they do not behave the same when used in baking. If you've ever kneaded bread dough or watched someone toss pizza dough, you've experienced the gluten protein. Gluten gives dough the elasticity it needs to be stretched and shaped. It's also what helps trap the gas released in the dough, which gives it its rise. Still, different varieties of wheat and its cousins act differently. Some, such

as soft wheat and rye, have lower amounts of protein, which means these flours aren't suitable to use by themselves in yeast breads (unless you desire a dense, brick-like loaf). Others, such as durum wheat, have high amounts of protein but contain a different gluten structure, making the flour not suitable for yeast breads but wonderful for pasta.

Early on, as I branched out into trying different flours, I was struck by the multitude of flavors. Kamut and einkorn wheat, for example, are a bit sweeter and milder, while spelt and emmer stand out in recipes with their distinct flavors. The subtle differences in the varieties can play a major role in the flavor of your final dish. My recommendation is to try a few different recipes for each grain and play around with the flours. Feel free to experiment and mix the flours in some recipes. One of my favorite flour combinations is barley, rye, and spelt.

KEEP YOUR GRAINS
SEPARATE

Because gluten-containing grains look extremely similar, create a system of labeling and storage. It's all too easy to mix the grains, and although there are differences, telling them apart can be difficult. When I buy grains from the bulk bins, I always write the name on the bag tie, and then I write it on masking tape for its glass storage jar.

BARLEY

Recently, barley has become one of my favorite grains to cook and grind into flour. Its flavor has a lovely nuttiness and sweetness that blends well with both sweet and savory dishes. Barley is one of my favorite grains to toss into salads, and it pairs particularly well with citrus dressings. Occasionally I will swap barley for rice when I make a hearty vegetable curry, as the sweetness of the barley is a great complement to the spiciness of the curry.

On top of the great taste, barley also packs into its tiny berry the highest fiber content of all the whole grains. Barley's nutrition is unique in that while the bran layer is packed with nutrients, the entire berry—the germ and endosperm—also contains good nutrients.

Even though barley's nutrients are distributed through the entire grain, it's best to look for hulled or hull-less barley instead of pearled barley. In hulled barley, the inedible outside layer is carefully removed to leave some of the bran layer intact, while the hull-less variety is grown without a tightly attached outer layer, making it easier to leave on the good bran layer. Pearled barley is not considered a whole grain because it is refined to remove the bran layer and thus is stripped of some of its nutrients.

Most barley sold in traditional supermarkets is pearled or in flakes similar to oats. For hulled or hull-less barley, check health food stores or places with bulk bins. To preserve nutrients in the flour, I always purchase hulled or hull-less. However, pearled barley and barley flakes are easier to process if you're using a spice mill or blender to mill the grains.

Barley Flour

The flavor of barley creates a wonderfully nutty and slightly sweet flour. Though barley flour contains gluten, its levels are such that it is often combined with traditional whole wheat flour because of its slightly gritty texture and lack of ability to create a rise in baked goods. When working with yeast breads, only one-quarter of the traditional flour should be substituted with barley flour to ensure that the bread will rise and have a light texture. However, I've found that certain items, such as cookies, pancakes, and even muffins, work really well with 100 percent barley flour.

Barley is best ground using an electric or hand grain mill but can be ground using a high-powered blender, food processor, or spice mill for a coarser grind. Immediately after grinding, barley flour is a bit airy, but it will settle after a day or two. For the most accurate measurement, it is best to go by weight.

Weights and Measurements

1 cup hulled barley = 180 g

1 cup barley flour = 120 g

1 cup (180 g) hulled barley = 1½ cups (180 g) barley flour

GRAPEFRUIT
BARLEY SCONES

These scones are a bit messier to make than traditional scones,
but I find the extra mess well worth the flavor. The tartness of the grapefruit is countered
nicely by the slight sweetness of the barley flour. However, if the tang of the grapefruit
isn't your favorite, trying subbing oranges instead.

YIELD: 4 LARGE OR 6 MEDIUM SCONES

1½ cups (180 g) barley flour

2 teaspoons baking powder

½ teaspoon sea salt

¼ teaspoon baking soda

1 medium grapefruit (275 to 300 g)

6 tablespoons (85 g) cold butter, cut into pieces

1 large egg, separated

¼ cup (60 ml) whole milk

3 tablespoons (60 g) honey, divided

Preheat the oven to 400°F (200°C, or gas mark 6). Line a baking sheet with parchment paper.

In a large bowl, whisk together the barley flour, baking powder, salt, and baking soda. Zest the grapefruit and rub the zest into the flour mixture. Carefully cut the peel off the grapefruit, dice, and squeeze the juice into a bowl. Set the juice and squeezed grapefruit pulp aside.

Cut the butter into the dry ingredients using a pastry blender, 2 knives, your hands, or a food processor until the dough is in pea-size pieces. In a smaller bowl, whisk together the egg white, milk, and 2 tablespoons (40 g) of the honey. Stir into the dry ingredients until the dough pulls together.

Transfer the dough from the bowl to a floured surface and pat into a rough 6 × 12-inch (15 × 30-cm) rectangle. Sprinkle the grapefruit pieces over and carefully roll into a log, jelly-roll style. Squeeze and pat the log into a slightly flatter log, about 1 inch (2.5 cm) thick.

Cut into 4 to 6 triangles or squares. Place on the baking sheet 1 to 2 inches (2.5 to 5 cm) apart. Whisk together the egg yolk, the remaining 1 tablespoon (20 g) honey, and 2 tablespoons (30 ml) of the reserved grapefruit juice; brush over the scones. Place the baking sheet in the oven and bake for 15 to 18 minutes, until the scones are golden and firm to the touch. Allow to rest for 10 minutes before transferring to a wire rack to cool. Store cooled scones in an airtight container at room temperature for 2 to 3 days, or freeze for later use.

DARK CHOCOLATE CHIP COOKIES

It's hard to improve on the traditional chocolate chip cookie, but the
barley flour steps these cookies up to the next level of flavor.
These cookies are slightly crispy on the outside with a deliciously soft
inside. I prefer to buy chunks of dark chocolate to chop,
but regular chocolate chips work as well.

YIELD: 18 TO 24 COOKIES

¾ cup (168 g) butter, softened

1 cup (192 g) loosely packed brown sugar

¼ cup (50 g) organic cane sugar

2 large eggs

1 teaspoon vanilla extract

¾ teaspoon baking powder

¾ teaspoon sea salt

¼ teaspoon baking soda

3 cups (360 g) barley flour

1½ cups (336 g) dark chocolate chunks

Preheat the oven to 375°F (190°C, or gas mark 5).

In the bowl of a stand mixer fitted with the paddle attachment or in a
large mixing bowl with a hand mixer, beat together the butter with both
sugars until blended and fluffy. Add the eggs, vanilla extract, baking
powder, salt, and baking soda, and continue to beat until everything is
well incorporated, scraping down the sides of the bowl as needed.

Add the flour and mix on low speed until the dough comes together and
everything is well incorporated. Stop the mixer and scrape down the sides
of the bowl again if necessary. Stir in the chocolate chunks.

Scoop the cookies onto baking sheets using an ice cream scoop or spoon,
using 2 to 3 tablespoons (56 to 84 g) of dough. Flatten slightly with the
palm of your hand and bake for about 12 minutes, or until the edges
are lightly brown; the centers will not be set but will continue to set as
they cool. Let cool for 10 minutes on the baking sheet and then transfer
to a wire rack to cool completely. Store in an airtight container at room
temperature for 3 to 4 days. If baking ahead of time, the cookies freeze
well. Just allow them to come to room temperature before serving.

Types of baking sheets can have a varied effect on the baking time of cookies. If
you're using a darker baking sheet, check the cookies after 10 minutes. If using
a silicone baking mat, such as a Silpat, these cookies may need a longer baking
time, closer to 15 minutes.

ASPARAGUS AND MOZZARELLA PIZZA WITH BARLEY CRACKER CRUST

I love a good thin-crust pizza, and using a cracker base ensures
a crispy crust without needing a pizza oven. The mozzarella is mild enough
not to overpower the asparagus, but goat cheese or Gorgonzola would be
wonderful on this pizza as well.

YIELD: ONE 10-INCH (25.5-CM) PIZZA

For the crust:

¾ cup (90 g) barley flour

¼ cup (30 g) soft whole wheat flour

2 tablespoons (16 g) cornstarch

1 teaspoon baking powder

½ teaspoon sea salt

¼ cup plus 2 tablespoons (90 ml) water

2 tablespoons (30 ml) olive oil

1 tablespoon (20 g) honey

For the toppings:

1 tablespoon (15 ml) olive oil

2 tablespoons (8 g) finely chopped
fresh dill

5 or 6 (80 g) medium stalks asparagus,
shaved with vegetable peeler or cut
into strips

2 scallions, diced

2 ounces (55 g) mozzarella, shredded

Preheat the oven to 450°F (230°C, or gas mark 8).

To make the crust: In a large bowl, combine the flours, cornstarch,
baking powder, and salt, and give it a good stir. Whisk together the
water, olive oil, and honey in a small bowl. Pour the wet ingredients into
the dry ingredients, adding a bit more water if the dough seems dry or
little more flour if the dough is too sticky. The dough should be easily
workable but not too sticky. Knead the dough 4 or 5 times.

On a piece of parchment paper sprinkled with 1 to 2 tablespoons
(7.5 to 15 g) barley flour, roll the dough out into a 12-inch (30-cm) circle.
Transfer the parchment paper to a baking sheet.

To make the toppings: Brush the olive oil over the crust and sprinkle on
the dill. Spread the asparagus and scallions over the crust. Finish with
the mozzarella.

Bake for 10 to 12 minutes, until the cheese is brown and the crust is
crisp. Remove from the oven and let sit for 2 to 3 minutes before cutting.

🖊 I am notorious for going easy on the cheese because I like all the ingredients to
shine. Add more cheese if desired.

RYE

I have always had a soft spot in my heart for rye, mainly because of my Swedish heritage. I grew up in a family where my grandfather would make silly Swede jokes about himself, and my grandmother would make Swedish rye bread that the entire family would fight over—my great-aunt loved it so much that she would cut the bread lengthwise to ensure a larger piece. I've only recently started exploring the possibilities of using rye outside of my grandmother's bread, but I've already decided it is a wonderful grain.

Rye's history isn't as old as many of the grains labeled "ancient," but cultivation of rye dates back to 400 BCE in Germany. The grain then spread throughout Europe and was used as peasant food while the nobility feasted on wheat, even though rye contains more nutrients than wheat. Rye's slow start can be partially attributed to its status as a weed growing among wheat plants. Nevertheless, rye began to gain popularity in countries that had poor soil because of its ability to grow in slightly harsh climates, such as that of Scandinavia.

Rye can be found in four forms: berries, cracked berries, flakes, and flour. The rye berries are the whole kernels with the inedible hull removed but with some of the nutritious bran layer left intact. The berry is slightly skinnier than a wheat berry and has a greenish tint to it. Cracked rye is a rye berry that has been processed into small pieces for quicker cooking, while rye flakes resemble old-fashioned oats.

Look for rye berries and flakes prepackaged or in the bulk bins of health food stores. I occasionally see rye flour in supermarkets but rarely the berries. The berries make a nutty addition to pilafs, and the rye flakes are great when combined with oat and barley flakes for a three-grain morning porridge.

Rye Flour

I find rye flour to be a uniquely flavored gluten-based flour. Rye berries have a slightly sour taste that is brought out in the flour. This flavor is most notable in pumpernickel bread, which is a dense, 100 percent rye bread.

Rye flour from the store can come in many different shades, depending on how much of the bran was processed out during milling. The lighter in color the flour appears, the less bran it has—which means less fiber. Milling rye from whole or cracked rye berries at home is a great choice because the final result is a dark flour, full of fiber. Rye flour doesn't contain as much gluten as wheat flour, and when used alone in baked goods, it creates an extremely dense texture. If a lighter texture is desired, used two-thirds rye flour combined with one-third wheat flour.

Rye berries are best ground in an electric or hand grain mill but can also be coarsely ground in a coffee grinder or high-powered blender. If using rye flakes, grind in a food processor and then sift for fine flour.

Weights and Measurements

1 cup rye berries = 180 g

1 cup rye flour = 120 g

1 cup (180 g) rye berries = 1½ cups (180 g) rye flour

SWEDISH RYE
QUICK BREAD

I wanted to include the recipe my grandmother used for yeasted
Swedish rye bread, but the end result when using 100 percent rye flour is a dense
pumpernickel loaf. So instead I opted to create a quick bread, in which a denser
texture is more expected, to mimic the flavors of her rye bread.

MAKES 8 TO 10 SLICES

2½ cups (300 g) rye flour

1½ teaspoons baking powder

½ teaspoon sea salt

¼ teaspoon baking soda

3 large eggs

½ cup (112 g) butter, melted

½ cup (112 g) plain whole-milk Greek
yogurt

½ cup (120 ml) whole milk

½ cup (176 g) sorghum sweetener

Preheat the oven to 375°F (190°C, or gas mark 5). Lightly butter or oil
a 5 × 8-inch (13 × 20-cm) bread pan.

In a large bowl, combine the rye flour, baking powder, sea salt, and
baking soda. In a separate bowl, whisk together the eggs, butter, yogurt,
milk, and sorghum. Pour the wet ingredients into the dry ingredients and
mix until well combined.

Pour the batter into the pan and spread evenly. Bake for 50 to 60 minutes, until the bread is firm to the touch and a knife inserted into the
center comes out clean. Let cool on a wire rack before removing from
the pan. Store the bread in an airtight container at room temperature
for 2 to 3 days, or store in the freezer.

🖊 Sorghum sweetener is produced from the green stalks of the same plant that
produces sorghum grains (page 122). The syrup has a rich taste, similar to
molasses, and can be found in most supermarkets next to the molasses and
honey. Molasses can be used in place of sorghum but the flavor of the bread
will be different.

RYE CREPES WITH HONEY APPLES AND GREEK YOGURT

One of my favorite desserts is a crepe, a little fruit, and a drizzle of
local honey. I am always amazed at the flavor that resides in such a simple dish.
Although sweet apples such as Gala and Cameo work best, I love Honeycrisp
with my crepes to add a hint of tartness.

YIELD: 6 TO 8 CREPES

For the crepes:

½ cup (60 g) rye flour

⅛ teaspoon sea salt

1 large egg

½ cup plus 2 tablespoons (150 ml) whole milk

1 tablespoon (14 g) butter, melted

2 teaspoons honey

For the honey apples:

3 apples (600 g), cored and cut into ¼-inch (6-mm) slices

¼ cup (85 g) honey

1 cup (224 g) whole-milk plain Greek yogurt

To make the crepes: In a large bowl, whisk together the rye flour, salt, egg, milk, butter, and honey until smooth.

Heat an 8-inch (20-cm) skillet over medium-low heat and lightly grease with oil. Working quickly, place a scant ¼ cup (60 ml) batter in the pan and tilt the pan in a circular motion so that the batter covers the entire pan. Cook the crepe for about 30 seconds, flip, and cook for another 15 seconds. Layer cooked crepes, slightly overlapping, on a plate.

To make the honey apples: Add the apples and honey to a large saucepan. Cook over medium-low heat, stirring occasionally, until soft, 4 to 5 minutes.

Spread the apples on one half of a crepe along with 1 to 2 tablespoons (15 to 30 g) Greek yogurt. Fold over to close the crepe, and spoon more apples and yogurt on top.

More often than not, I'm making these crepes just for my husband and me. Because we cannot eat a full batch, I will make about half the filling and freeze any extra crepes for later use. To freeze, separate the crepes with waxed paper and place in a freezer-safe container. Crepes thaw quickly when left out on the counter and make fast work of dessert.

French Toast with Rye Bread and Bourbon Cherries

Making french toast with quick bread is a bit indulgent,
but some weekend mornings call for a little more of an extravagant breakfast.
The bourbon in the topping is optional—you can add a pinch of cinnamon instead.

YIELD: 2 SERVINGS

For the french toast:

2 large eggs

2 tablespoons (30 ml) heavy cream

1 tablespoon (14 g) coconut oil

4 slices (1 inch or 2.5 cm each) Swedish
Rye Quick Bread (page 32)

For the cherries:

2 cups (360 g) cherries, pitted

4 teaspoons (20 ml) maple syrup

1 tablespoon (15 ml) bourbon (optional)

To make the french toast: Whisk together the eggs and heavy cream.
In a large skillet or griddle over medium heat, melt the coconut oil.

Soak each piece of bread in the egg mixture until coated on both sides
and place in the skillet. Cook, turning once, until the French toast is
browning and crisp, 3 to 5 minutes.

To make the cherries: Combine the cherries and maple syrup in a small
saucepan. Bring to a boil, reduce to a simmer, and cook until the cherries
are soft and the liquid has reduced, 4 to 6 minutes. Stir in the bourbon,
if using.

Serve the french toast with the cherries.

KAMUT
(KHORASAN WHEAT)

When I first joined my local CSA, I viewed my weekly haul of vegetables as a challenge to learn about and experiment cooking with the different items. I have my CSA to thank for my love of swiss chard, kohlrabi, and fava beans. I look at the bulk bins in health food stores in a similar light. I'm drawn to the different shapes and colors of all the grains, and often I'm on my smartphone looking up what something might be. This is how I discovered kamut.

Of all the grains, kamut has the most interesting history. Although the exact history of the grain isn't known, we do know that it originated in ancient Egypt and in more recent times was brought to the United States by a U.S. airman in the 1940s. The grain didn't take off right away; it took until the 1970s, when an agriculture science professor who remembered seeing it started growing and selling kamut.

Kamut is the trademarked name for khorasan wheat, and the branding ensures quality of the grain, specifically that it is organic, 99 percent free of modern wheat, and has a protein range between 12 and 18 percent. The grain itself is similar in color to modern wheat but is twice the size. The protein level of kamut is also higher than that of modern wheat, making it a good substitute for wheat berries.

Because this grain is still gaining in popularity, it can be a bit tougher to find compared with other wheat varieties. Look in the bulk bins of specialty stores or health food stores. Kamut can sometimes be found in the prepackaged grain aisles of large supermarkets.

Kamut Flour

When I first starting baking with kamut flour, I was greatly surprised. I expected that it would taste like traditional wheat flour. However, kamut has a smoother and sweeter taste that really shines through in the flour. This flour makes a wonderful replacement for wheat, especially in baked goods. I've found that because of the bit of sweetness of the flour, not as much sugar is needed in the recipe.

Kamut is best ground in an electric or hand grain mill but can also be milled in a high-speed blender or coffee grinder for small batches. I recommend grinding the flour on a pastry setting, if your grain mill has that option. Using the finer flour keeps baked goods slightly airier and not as dense as some items made with wheat flour.

Kamut flour absorbs more moisture than modern wheat flour, and typically recipes call for ¾ cup of kamut to replace 1 cup of wheat flour. However, I've found that it's best to use 1 to 2 tablespoons (7.5 to 15 g) less per 1 cup (120 g) wheat flour or to use slightly more liquid, such as milk or melted butter, to keep baked goods moist.

Weights and Measurements

1 cup kamut berries = 180 g

1 cup kamut flour = 120 g

1 cup (180 g) kamut berries = 1½ cups (180 g) kamut flour

Kamut English Muffins with Quick Blueberry Ginger Jam

I rotate between three breakfast items throughout the week:
yogurt, eggs, and these English muffins smothered with peanut butter and jam.
I'll make a double batch and freeze them all, pulling
one or two out at a time to thaw.

YIELD: 10 ENGLISH MUFFINS AND 1 CUP (320 G) JAM

For the English muffins:

1 cup (235 ml) buttermilk

1 package (2¼ teaspoons) active dry yeast

3 tablespoons (60 g) honey

2½ to 2¾ cups (300 to 330 g) kamut flour, divided

1 tablespoon (14 g) butter, melted

1 teaspoon sea salt

Cornmeal, for dusting

For the jam:

2½ cups (12 ounces, or 340 g) blueberries

2 tablespoons (40 g) honey

1 tablespoon (15 ml) fresh lime juice

1 teaspoon grated fresh ginger

The jam will keep for up to 1 week in the refrigerator in an airtight container.

To make the English muffins: In a small skillet, heat the buttermilk until warm but not scalding (110°F to 120°F, or 43°C to 49°C). In the bowl of a stand mixer fitted with the dough hook, combine the warm buttermilk, yeast, and honey. Let sit until the yeast becomes active and frothy, 5 to 10 minutes.

Stir in 2 cups of the flour, the butter, and salt. Begin to knead the dough on the lowest speed. Add additional flour, ¼ cup (30 g) at a time, until the dough begins to pull away from the side of the bowl but is still slightly sticky.

Remove from the mixer and shape the dough into a ball. Place in a lightly oiled bowl, cover with a damp towel, and place in a draft-free area. Let rise until nearly doubled in size, about 1½ hours.

Once the first rise is complete, carefully pat out into a ¼- to ⅓-inch (6-mm to 1-cm) thickness on a cornmeal-dusted surface. Using a 3-inch (7.5-cm) biscuit cutter or the end of a glass that is approximately the same diameter, cut out English muffins and place on a lightly greased baking sheet. Gather scraps and shape into 1 or 2 more disks. Cover again with a slightly damp towel and let rise for another 45 minutes.

Heat a griddle or large skillet over medium heat and preheat the oven to 400°F (200°C, or gas mark 6). Brown the English muffins on each side in the griddle, 2 to 3 minutes, and place back on the baking sheet. Bake for 10 to 12 minutes, until firm. Let cool slightly before serving.

To make the jam: Combine all of the ingredients in a small saucepan over medium heat. As the blueberries warm, smash them with the back of a wooden spoon. Bring to a boil, stirring often, and continue to cook until the jam thickens, 10 to 12 minutes. Place in the refrigerator to let cool and thicken further.

SWEET CORN AND RICOTTA RAVIOLI WITH BUTTER BASIL SAUCE

When I was growing up in Illinois, summers were always about
scouting out the best farmer to buy sweet corn from and eating as much of it as we could
in July. The kamut flour's subtle flavor pairs nicely with the ricotta and sweet corn filling.

YIELD: 4 SERVINGS (18 TO 20 LARGE RAVIOLI)

For the dough:

2 cups (240 g) kamut flour
½ teaspoon sea salt
3 large eggs

For the filling:

1 tablespoon (15 ml) olive oil
½ medium red onion, diced
Kernels from 1 large ear corn, or 1 cup
  (163 g) frozen corn kernels, thawed
¾ cup (180 g) whole-milk ricotta
½ teaspoon sea salt
½ teaspoon freshly ground black pepper

For the sauce:

¼ cup (56 g) butter
2 tablespoons (30 ml) olive oil
1 clove garlic, minced
½ cup (20 g) loosely packed fresh basil,
  julienned

🖊 If you don't have a ravioli stamp or cutter,
simply use a knife to cut the pasta into
squares and seal the edges by pressing down
with a fork.

To make the dough: Combine the kamut flour and salt on a clean, flat surface. Make a well in the middle and crack the eggs in the center. Using a fork, whisk the eggs, and then slowly begin to incorporate the flour. Continue to combine the flour and eggs until a paste-like texture forms. Keep mixing, eventually trading the fork for your hands, and knead the dough into a smooth ball. Let rest while making the filling.

To make the filling: In a large skillet over medium-low heat, heat the olive oil. Add the onion to the skillet and sauté until translucent, 6 to 7 minutes. Stir in the corn kernels and continue to cook until the corn is tender and beginning to char, 8 to 10 minutes. Remove from the heat and transfer to a medium bowl. Gently add the ricotta, salt, and pepper. Set aside.

Bring a pot of salted water to a boil. Divide the dough into 8 pieces. Using the pasta attachment for a stand mixer or a rolling pin, roll out 1 piece of pasta to a ⅛-inch (3 mm)-thick strip that's about the width of a ravioli stamp. (Cover the rest of the dough with a damp cloth.) The pasta should be thin but still hold together. Repeat with remaining pieces.

Lay 4 strips of dough horizontally on a work surface. Spoon the ricotta mixture by the tablespoonful (16 g) in even spacing onto each strip, 5 or 6 spoonfuls per strip. Place another strip of dough on top and cut into squares using a square ravioli stamp or cutter. Add 4 or 5 ravioli to the boiling water, and cook until the ravioli float to the top, 5 to 6 minutes. Remove with a slotted spoon, place in a serving bowl, and continue with the remaining ravioli.

To make the sauce: Melt the butter in a small saucepan over medium-low heat and add the olive oil. Stir the garlic into the butter mixture along with the basil, and cook until the garlic is fragrant. Toss with the ravioli and serve.

UPSIDE-DOWN PEAR
KAMUT CAKE

My first bite of this cake completely surprised me. Sometimes
wheat flavor can be overpowering, especially in desserts, but the combination of
cinnamon, cardamom, maple syrup, and pears plays just right with the sweet
flavor of kamut flour. I recommend serving this cake with a scoop of
vanilla ice cream or a drizzle of heavy cream.

YIELD: 6 TO 8 SERVINGS

¼ cup (60 ml) plus ½ cup (120 ml)
maple syrup, divided

1 or 2 (240 g) Bosc pears, cored and
cut into ¼-inch (6-mm) slices

1½ cups (180 g) kamut flour

1 teaspoon baking powder

1 teaspoon ground cinnamon

½ teaspoon ground cardamom

½ teaspoon sea salt

¼ teaspoon baking soda

2 large eggs

½ cup (112 g) whole-milk plain Greek
yogurt

⅓ cup (74 g) butter, melted and slightly
cooled

1 teaspoon vanilla extract

Preheat the oven to 375°F (190°C, or gas mark 5).

In an 8-inch (20-cm) cast-iron or other ovenproof skillet over medium-low heat, heat ¼ cup (60 ml) of the maple syrup. Overlap the pears in a circular pattern, starting in the center and fanning out. Cook for 2 to 3 minutes, until the pears are soft and the maple syrup is warm. Remove from the heat.

In a large bowl, stir together the kamut flour, baking powder, cinnamon, cardamom, salt, and baking soda. In a separate bowl, whisk together the eggs, yogurt, butter, remaining maple syrup, and vanilla. Pour the wet ingredients into the dry ingredients and stir until combined.

Pour the batter over the pears and smooth out to cover. Bake for 22 to 24 minutes, until a knife inserted into the center comes out clean. Remove from the oven and let cool slightly. Run a knife along the edge and lay a serving plate on top of the cake. Flip the entire skillet and serving plate over so that the cake falls out of the pan and onto the plate. Remove the pan for serving. Store in an airtight container at room temperature for 2 to 3 days. If the cake dries out slightly after storing, serve with heavy cream.

I typically use Bosc pears because of their firmness and ability to hold their shape when cooked. However, depending on the season, I will use other pears based on their availability.

EMMER, EINKORN, AND SPELT

Emmer, einkorn, and spelt each have a long, rich history of production and consumption, but they are often confused for one another by being called by another name: farro. In Italy, emmer, einkorn, and spelt are *farro medio*, *farro piccolo*, and *farro grande*, respectively. Although the majority of grains sold in the United States under the farro name are actually emmer, there can still be some confusion. When purchasing "farro," look for specifics such as:

Emmer/*Triticum dicoccum*/*farro medio*

Einkorn/*Triticum monococcum*/*farro piccolo*

Spelt/*Triticum spelta*/*farro grande*

Each grain has specific flavors and textures that emerge when used both whole in cooking and in grinding to use as flour, which means they do not always substitute well for one another.

Emmer (*Farro Medio*)

Emmer has made a large impression in my kitchen. When I first tried the whole grain, I fell for the slightly sweet, nutty taste and soft texture that, when compared with other wheat grains, had a chewier texture. I love using emmer in place of traditional rice in risottos, when making whole-grain salads, and even as a breakfast porridge with fresh berries and a bit of honey.

Emmer is grown in the mountainous regions of Asia and Europe, with a substantial history in Italy. Emmer can be found in health food stores and even some regular grocery stores as the grain is becoming more popular. Often the grain sold in stores is semi-pearled, meaning that part of

the bran and nutrients have been removed. This is great for a quicker cooking time, but I always look for hulled, whole-grain emmer, which retains more nutrients.

Emmer Flour

Emmer flour is high in protein and fiber, making it a wonderful substitute for modern wheat flour. The taste is mild, with a hint of sweetness, similar to kamut. Whole-grain emmer flour does have a coarser texture compared with other wheat flours. The coarser texture doesn't work as well in pastries but makes a wonderful addition to breads. If I'm working with pastries, I grind the semi-pearled emmer. With part of the bran removed, it makes for a lighter flour that works well in pasties such as the Rosemary Sweet Potato Hand Pies on page 43. Start with the whole grain and move to the semi-pearled only if you find the flour too strong in flavor or tricky to work with.

Grinding emmer is best done in a grain mill or high-powered blender. Small batches can be ground in a coffee grinder and sifted to create the flour. I often substitute emmer for traditional wheat when baking, especially in pita, such as the recipe on page 45, or in place of the kamut in the english muffins on page 37. I also find that emmer flour makes for great pasta, especially in lasagna.

Weights and Measurements

1 cup emmer, whole or pearled = 180 g

1 cup emmer flour = 120 g

1 cup (180 g) emmer = 1½ cups (180 g) emmer flour

ROSEMARY SWEET POTATO HAND PIES

Hand pies are the perfect accompaniment to a long hike or picnic.
These pies can be baked ahead of time and taken anywhere. If Gorgonzola is too strong
of a cheese for you, try using a shredded cheese such as mozzarella or white Cheddar.

YIELD: 8 HAND PIES

For the dough:

1½ cups (180 g) emmer flour, preferably pearled (see Note)

1 tablespoon (13 g) organic cane sugar

½ teaspoon sea salt

½ cup (112 g) cold butter, cut into pieces

2 to 3 tablespoons (30 to 45 ml) water

For the filling:

2 cups (275 g) cubed sweet potato (about 1 large or 2 medium)

2 tablespoons (30 ml) olive oil, divided

2 ounces (55 g) Gorgonzola, crumbled

2 tablespoons (3.4 g) chopped fresh rosemary

Preheat the oven to 400°F (200°C, or gas mark 6).

To make the dough: In a food processor or a medium bowl, combine the flour, sugar, and salt. Pulse or cut in the butter using a pastry blender, 2 knives, or your hands. Once the dough is in pea-size pieces, pulse in 1 tablespoon (15 ml) water at a time until the dough begins to come together. Remove from the food processor and shape into a disk without handling the dough too much. Wrap in plastic wrap and place in the refrigerator while the sweet potatoes bake.

To make the filling: Toss the sweet potatoes with 1 tablespoon olive oil. Spread on a baking sheet and bake until tender, 25 to 30 minutes. Combine in a bowl with the Gorgonzola and rosemary. Lightly mash the sweet potatoes into the cheese. Let cool completely.

Remove the dough from the refrigerator when the sweet potatoes are cool and let sit for 5 to 10 minutes, until the dough is easily workable. Divide the dough in half and, on a floured surface, roll 1 half into a square about ¼ inch (6 mm) thick. Cut the square into 4 smaller squares and scoop 2 to 3 tablespoons (21 to 42 g) filling into the center of each. Fold the dough over the filling, forming a triangle, and with a fork, press down on the edges to seal. Repeat with the remaining dough.

Place the hand pies on a baking sheet. Brush with the remaining 1 tablespoon (15 ml) olive oil and bake for 25 minutes, or until the crust is brown and crisp. Let cool slightly.

Using whole-grain emmer instead of pearled can have an effect on your dough workability. If using whole-grain emmer, be patient, as the dough may tend to fall apart and crumble slightly.

PITA BREAD WITH EMMER FLOUR

There is never a time when I do not have pita bread stashed
in my freezer. I love using it for a quick sandwich, personal pizza, or snack
with hummus. Of course, I always like to make sure my pitas have the perfect
pocket, which occasionally can be tough to achieve. To do this, make sure
the dough is soft and avoid rolling the pitas too thin.

YIELD: 8 TO 10 PITA ROUNDS

1 package (2¼ teaspoons) active dry yeast

1 cup (235 ml) warm water (105°F to
110°F, or 40°C to 43°C)

2 tablespoons (40 g) honey

3 to 3½ cups (360 to 420 g) emmer flour,
divided

3 tablespoons (45 ml) olive oil

1 teaspoon sea salt

Combine the yeast, water, and honey in the bowl of a stand mixer with
the dough hook attachment. Let the mixture sit for 5 to 10 minutes, until
the yeast becomes active and frothy. Stir in 2 cups (240 g) of the emmer
flour, the olive oil, and salt.

Start the mixer on the lowest speed and add 1 more cup (120 g) of flour.
Knead for 5 minutes, adding extra flour, 1 to 2 tablespoons (7.5 to 15 g)
at a time, as needed. The dough should pull away from the sides of the
mixer and be slightly sticky but easy to handle. Shape into a ball, place
in a lightly oiled bowl, cover with a damp towel, and place in a warm,
draft-free area. Let rise for 1½ hours.

Preheat the oven to 500°F (250°C, or gas mark 10) with a pizza stone on
the lowest rack.

Divide the dough into 8 to 10 pieces. On a floured surface, roll each
piece into a ¼-inch (6 mm)-thick circle, covering with a damp towel after
rolling each piece.

Place 2 pita rounds on the pizza stone at a time and bake for 4 to
5 minutes. The bread should puff up and be slightly golden. Remove and
cover the warm rounds with a damp towel to help keep them soft. Repeat
with the remaining pita rounds. Once the pitas have cooled, freeze in an
airtight container. Thaw as needed.

Pita dough should be soft and easily workable but still slightly sticky. If the
dough is still too sticky when rolling out the rounds, work in a little extra flour.

TOMATO COBBLER WITH CHIVE EMMER BISCUIT TOPPING

One of my favorite indulgences during the summer is a rich
berry cobbler smothered with homemade vanilla ice cream. This tomato version
is its savory counterpart. The filling highlights juicy summer tomatoes, which
pair perfectly with the slightly rustic taste of the emmer flour biscuits.

YIELD: 4 SERVINGS

For the filling:

1 tablespoon (15 ml) olive oil

1 small (100 g) red onion, minced

1 clove garlic, minced

1 pound (455 g) grape tomatoes

2 teaspoons dried thyme

½ teaspoon freshly ground black pepper

¼ teaspoon sea salt

For the biscuits:

1 cup (120 g) emmer flour

2 teaspoons baking powder

¼ teaspoon sea salt

2 tablespoons (28 g) cold butter, cut into
pieces

1½ ounces (43 g) cold cream cheese, cut
into pieces

3 tablespoons (9 g) minced fresh chives

3 tablespoons (45 ml) buttermilk

2 teaspoons honey

1 tablespoon (15 ml) heavy cream

Preheat the oven to 375°F (190°C, or gas mark 5).

To make the filling: In an 8-inch (20-cm) ovenproof skillet over medium heat, heat the olive oil. Add the onion and cook until starting to brown, about 10 minutes. Add the garlic and cook for 1 minute more.

Turn off the heat and add the tomatoes, thyme, pepper, and salt. Place the skillet in the oven and bake for 25 minutes. Make the biscuits.

To make the biscuits: Combine the flour, baking powder, and salt, and whisk together. Cut the butter into the dry ingredients, using your hands or a food processor, until the butter is in smaller pieces. Add the cream cheese and continue to cut in (or pulse) until both the cream cheese and butter are in pea-size pieces. Stir in the chives.

Whisk together the buttermilk and honey and pour over the dry ingredients, stirring until the dough pulls together. For a rustic look, drop the biscuit dough by the spoonful over the cooked tomato mixture. For a more refined look, roll the dough out on a floured surface, cut into 4 biscuits, and arrange on top of the tomatoes. Brush the tops of the biscuits with the cream.

Return the skillet to the oven and bake for another 20 minutes, or until the biscuits are firm to the touch and the tomato mixture is bubbling.

If you don't own a cast-iron skillet, sauté the onions and garlic and then transfer them to an 8-inch (20-cm) round baking pan. Continue with the remainder of the recipe as directed.

EINKORN
(*FARRO PICCOLO*)

Keeping track of all the different kinds of wheat can be a bit overwhelming. At first, I thought all the wheat berries and flours were similar, but this notion led to a lot of failed baking experiments. Each variety of wheat berry has a different level of gluten and a unique makeup of nutrients and acts differently in recipes. It may take a bit of time to figure out how best to use this grain, but that extra time is well worth it. The mild wheat flavor and light flour makes it perfect for almost anything.

Einkorn, from the German for "single grain," is one of the earliest cultivated forms of wheat that has remained unchanged throughout the thousands of years it has been grown. Modern wheat has been hybridized to make new wheat grains. For example, kamut and durum wheat both descend from emmer wheat. Because einkorn hasn't changed over the many years, the grain has a different gluten structure than other varieties of wheat and contains more protein.

The grain is aptly called *farro piccolo* in Italy because of its miniature size compared with emmer and spelt. The grain has slowly been increasing in popularity as a specialty wheat but is still grown in only a few select mountainous regions of Europe. Because of this, einkorn is sold in very few retail outlets. While I have been able to find the flour and sometimes the berries in health food stores, there are a couple of companies online that you can order from and have einkorn shipped to your door.

Einkorn Flour

Because of einkorn's gluten structure, my first few baking experiences with the flour left me baffled. I was making bread in the stand mixer and patiently waiting for the dough to pull away from the sides of the bowl while continuing to add flour. After using a few cups more than what the recipe called for, I decided to let the dough rise and try baking the bread. The end result was a hard brick—and only myself to blame.

Einkorn flour absorbs water slowly, and with dough that requires time to rest, it is best to leave the dough slightly wet. While the dough rests, the flour will absorb the water, making it perfectly workable.

Einkorn flour grinds best in a grain mill or high-powdered blender. It is a sweet, lighter flour that works well in any recipe calling for wheat flour, especially pastries. If substituting einkorn flour for wheat flour in a bread recipe, increase the liquid or lower the amount of flour used to avoid problems. For quick breads and pancakes, use 1 to 2 tablespoons (6 to 12.5 g) less flour and 1 to 2 tablespoons (15 to 30 ml) more liquid per cup (235 ml) called for in the recipe.

Weights and Measurements

1 cup einkorn berries = 180 g

1 cup einkorn flour = 100 g

1 cup (180 g) emmer = 1¾ cups (175 g) emmer flour

EINKORN FLOUR
HAMBURGER BUNS

There are certain recipes that I feel everyone should have in their
back pocket, and hamburger buns fit right into that category. At the start of
grilling season, I make up a big batch of buns to freeze and use throughout the
ensuing weeks. The lightness of the einkorn flour creates a lovely texture
that makes the buns perfect for any cookout.

YIELD: 12 BUNS

1 cup (235 ml) warm water (105°F to 110°F, or 40°C to 43°C)

1 package (2¼ teaspoons) active dry yeast

¼ cup (85 g) honey

6 cups (600 g) einkorn flour, divided

3 large eggs, divided

¼ cup (56 g) butter, melted

1 teaspoon sea salt

1 tablespoon (15 ml) water

1 to 2 tablespoons (8 to 16 g) sesame and/or poppy seeds

In a large bowl, stir together the water, yeast, and honey. Let sit for 5 to 10 minutes while the yeast activates and becomes frothy. Add 4 cups (400 g) of the flour, 2 of the eggs, the butter, and salt. Stir until well combined. Continue to stir and add flour, ¼ cup (25 g) at a time, incorporating the flour after each addition. Switch to kneading with your hands, covered with flour, once the dough is tough to stir. Knead the dough a few times until all the flour is incorporated. The dough will still look wet, but resist the urge to add more flour. Place in a bowl, set in a warm, draft-free area, and cover with a damp towel. Let rise for 1½ hours.

Scrape the dough onto a floured surface and divide into 12 equal pieces. Roll each piece into a ball, using additional flour if the dough is sticking. Place 1 piece on a baking sheet lined with parchment paper and press down to form a circle that is roughly ½ inch (1.2 cm) thick and 2½ to 3 inches (6.2 to 7.5 cm) wide. Repeat with the remaining dough pieces. Cover again and let rise for 1 hour.

Preheat the oven to 400°F (200°C, or gas mark 6).

Whisk together the remaining egg and the 1 tablespoon (15 ml) water, then lightly brush the mixture over the tops of the rolls. Sprinkle with the seeds. Bake for 10 to 12 minutes, until the rolls are puffed and golden. Let cool before slicing.

Because of einkorn's slow water absorption rate, knead the dough by hand and not in a stand mixer. Using a stand mixer will overwork the dough and you will want to add more flour than needed because the dough will look too soft and sticky.

BISCUIT CINNAMON ROLLS

In the town where I was raised, there is a restaurant known for
its cinnamon rolls. For the longest time I was baffled as to how the cinnamon rolls
were so perfectly flaky until I did a bit of research and realized that the base was not
a yeasted dough, but biscuit mix. These cinnamon rolls take me back home and
are a bit less time-consuming than their yeasted counterpart.

YIELD: 8 CINNAMON ROLLS

For the dough:

3 cups (300 g) einkorn flour

1 tablespoon (13.8 g) baking powder

½ teaspoon sea salt

6 tablespoons (85 g) cold butter, cut into pieces

½ cup (120 ml) buttermilk

1 large egg

1 tablespoon (20 g) honey

For the filling:

¼ cup (56 g) plus 2 tablespoons (28 g) butter, melted and divided

¼ cup (60 g) plus 2 tablespoons (30 g) packed brown sugar, divided

3 teaspoons ground cinnamon, divided

Preheat the oven to 425°F (220°C, or gas mark 7). Line a baking sheet with parchment paper.

To make the dough: In a large bowl, stir together the einkorn flour, baking powder, and salt. Cut in the butter using 2 knives or your fingers until the dough is in pea-size pieces. In a small bowl, whisk together the buttermilk, egg, and honey, then pour into the dry ingredients. Stir the mixture until the dough comes together.

To make the filling and assemble: Scoop the dough onto a floured surface and pat into roughly an 8 × 12-inch (20 × 30-cm) rectangle. Brush on ¼ cup (56 g) of the melted butter. Combine ¼ cup (60 g) of the brown sugar with 2 teaspoons of the cinnamon and sprinkle over the butter. Roll into a log, starting with the long edge closest to you. Lightly squeeze the roll into a log. Brush with the remaining 2 tablespoons (28 g) melted butter and sprinkle the remaining 2 tablespoons (30 g) brown sugar and 1 teaspoon cinnamon over the top. Cut into 8 equal pieces and transfer to the baking sheet.

Bake for 18 to 20 minutes, until the top is brown and crisp. Remove and let cool slightly before serving. These rolls are best served fresh but can be stored in an airtight container at room temperature for 2 to 3 days.

LEMON HONEY BARS

I find store-bought lemon bars to be too sweet. I prefer to
showcase the tartness of the lemon in this custardy dessert. However, if you
find these bars too tart, add an additional ¼ cup (85 g) honey to the filling
to help cut some of the tartness. These bars also are delicious
made with lime juice or orange juice.

YIELD: 16 SMALL OR 9 LARGE SQUARES

For the crust:
1¼ cups (125 g) einkorn flour
¼ cup (56 g) butter, melted
2 tablespoons (40 g) honey

For the filling:
4 large eggs
½ cup (170 g) honey
⅓ cup (80 ml) fresh lemon juice
⅓ cup (80 ml) water
1 tablespoon (6 g) lemon zest
½ teaspoon vanilla extract
¼ cup (25 g) einkorn flour

Preheat the oven to 350°F (180°C, or gas mark 4). Lightly grease an
8 × 8-inch (20 × 20-cm) baking pan.

To make the crust: In a medium bowl, combine the flour, butter, and
honey and stir to form a dough. Remove from the bowl and pat the
dough into the bottom of the prepared pan. Bake for 15 to 20 minutes,
until the crust is golden. Let cool for 10 to 15 minutes.

To make the filling: Combine the eggs, honey, lemon juice, water, lemon
zest, vanilla, and flour in a blender. Puree until smooth, and pour onto
the cooled crust. Return the pan to the oven and bake until the filling
is set, 20 to 22 minutes. Let cool before slicing. Store in an airtight container at room temperature for 3 to 4 days.

SPELT
(FARRO GRANDE)

Spelt, as a grain, never really caught my attention. I'd see spelt flour incorporated into multigrain loaves of bread and in pasta, but that was my extent of interaction with spelt. When I started grinding my own flours for bread, I purchased spelt berries for a multigrain mix. Before I ground the berries into flour, I cooked a few and found the taste to be pleasantly nutty and hearty, with a chewy texture. Since then, I've always had a jar of spelt berries sitting around for the occasional spelt berry salad or multigrain bread.

Spelt's origins are a bit hazy; in fact, it has two potential lineages: one in Central Europe and one in Asia. Spelt is known for having a tough outer husk and a lower gluten content than modern wheat. The spelt berry is called *farro grande*, in Italy, as the berry is larger than the modern wheat berry and can be identified by its long point. Spelt is grown as a specialty grain and can be found prepackaged or in the bulk bins of health food stores.

Spelt Flour

Even though spelt is a cousin of wheat, the flour acts a bit differently. Spelt flour is more water soluble, which means less liquid is often needed in recipes. When adapting a recipe using spelt flour, start with 2 to 3 tablespoons (30 to 45 ml) less liquid. I stand by the saying, "You can always add more liquid; you can't take it out." I also find your best judgment to be a great guide—if the dough or batter looks too dry, add a bit more liquid.

While yeast bread can be made from the lower-gluten spelt flour, kneading time is shorter because there is no need to strengthen the gluten. Over-kneading spelt doughs can result in a crumbly texture.

Spelt can be ground in a similar way as the other wheat varieties. It works best in a grain mill but can also be done in a high-powered blender or coffee grinder. Often in stores, both spelt and white spelt flour are sold. White spelt flour has been refined and is not considered a whole-grain flour. Grinding spelt at home is a great way to ensure you are getting all the nutrients and fiber in your flour.

Weights and Measurements

1 cup spelt berries = 180 g

1 cup spelt flour = 120 g

1 cup (180 g) spelt berries = 1½ cups (180 g) spelt flour

CHEDDAR ROSEMARY SPELT SCONES

Scones are my favorite pastry, and although the sweet ones
are usually what I eat, the savory scone can be a wonderful substitute
for dinner rolls or enjoyed as an afternoon snack. The combinations of
cheese and herbs are endless, but I'm quite smitten with
Cheddar and rosemary.

YIELD: 6 TO 8 SCONES

1½ cups (180 g) spelt flour

2 tablespoons (26 g) organic cane sugar

2¼ teaspoons baking powder

½ teaspoon sea salt

¼ teaspoon baking soda

6 tablespoons (84 g) cold butter, cut into pieces

1½ cups (150 g) shredded Cheddar, divided

2 tablespoons (3.4 g) minced fresh rosemary

1 egg yolk

¼ cup plus 2 tablespoons (90 ml) buttermilk

2 tablespoons (30 ml) heavy cream

Preheat the oven to 425°F (220°C, or gas mark 7). Lightly grease a baking sheet or line it with parchment paper.

In a large bowl, stir together the spelt flour, sugar, baking powder, salt, and baking soda. Cut in the butter using a pastry blender, 2 knives, or your hands until the dough is in pea-size pieces. Stir in 1¼ cups (125 g) of the cheese and the rosemary. In a smaller bowl, whisk together the egg yolk and buttermilk and then stir them into the dry ingredients until the dough pulls together.

Scoop the dough onto a floured surface. Pat the dough into a circle ½ inch (1.2 cm) thick and cut into 6 to 8 triangles. Place 1 to 2 inches (2.5 to 5 cm) apart on the prepared baking sheet. Brush the heavy cream over the scones and sprinkle the remaining ¼ cup (25 g) cheese over the top. Bake for 15 to 18 minutes, until the scones are golden and firm to the touch. Let cool on the baking sheet for 5 minutes, then transfer to a wire rack to finish cooling. Store in an airtight container at room temperature for 2 to 3 days, or freeze for later use.

Spelt Pull-Apart Dinner Rolls

Even though yeasted spelt doughs are a bit finicky,
I love making these rolls. The spelt sets up a perfect blank canvas that can
hold many different flavor combinations. When brushing the olive oil on the rolls,
sprinkle on your favorite herb combination, grated cheese, or minced garlic
to bake in the pull-apart sections.

YIELD: 9 TO 12 ROLLS

¾ cup (176 ml) warm water (105°F to 110°F, or 40°C to 43°C)

1 package (2¼ teaspoons) active dry yeast

2 tablespoons (40 g) honey

2½ cups (300 g) spelt flour, divided

2 tablespoons (30 ml) olive oil

½ teaspoon sea salt

2 tablespoons (30 ml) olive oil or melted butter

1 large egg

1 tablespoon (15 ml) water

Combine the warm water, yeast, and honey in the bowl of a stand mixer fitted with the dough hook and let sit until the yeast becomes active and frothy, 5 to 10 minutes. Stir in 2 cups (240 g) of the flour, the olive oil, and the salt.

Start the stand mixer on low speed and continue to add flour, 2 to 3 tablespoons (15 to 22.5 g) at a time. Continue to knead the dough and add flour until the dough pulls away from the side of the bowl.

Remove the dough from the mixer and form into a ball. Place in a lightly oiled bowl, cover with a damp cloth, and place in a warm spot. Let rise until doubled in size, 1 to 1½ hours.

After the first rise, turn the dough out onto a floured surface and roll out into an 8 × 12-inch (20 × 30-cm) rectangle. Brush the olive oil or melted butter over it. Cut the dough across the shorter length into 6 strips (it's easier to handle). Stack the strips on top of each other and cut through them, creating 9 to 12 stacks. Brush the wells of a standard muffin pan with oil and place each stack in a well. Cover and let rise again for 1 hour.

Preheat the oven to 400°F (200°C, or gas mark 6).

In a small bowl, whisk together the egg and water. Brush the tops of the rolls with the egg wash and bake for 15 to 18 minutes, or until the rolls are golden. Let cool for 5 minutes before removing from the muffin pan. Store in an airtight container at room temperature for 1 to 2 days, or freeze for later use.

ZUCCHINI AND CORN EMPANADAS

These vegetarian empanadas hit the spot when I'm craving
street fair–style food without all the grease. Divide the dough into even
smaller portions and make cute finger food that's perfect for any party!
Serve with your favorite salsa, sour cream, or plain Greek yogurt

YIELD: 12 EMPANADAS

For the dough:

2¼ cups (270 g) spelt flour

1½ teaspoons sea salt

10 tablespoons (140 g) cold butter, cubed

1 large egg

2 tablespoons (30 ml) heavy cream

For the filling:

1 tablespoon (15 ml) olive oil

1 small red onion, minced

2 small jalapeño chile peppers, minced

1 medium zucchini (250 g), diced

Kernels from 2 medium ears corn, or
   1¾ cups (300 g) frozen corn kernels,
   thawed

½ cup (8 g) chopped fresh cilantro

½ teaspoon sea salt

¼ teaspoon crushed red pepper

Juice of 1 lime

For the assembly:

1 large egg

1 tablespoon (15 ml) water

Paprika, for sprinkling

To make the dough: In a medium bowl, combine the flour and salt. Cut the butter into the flour mixture with either a pastry blender or your hands (my preferred method) until the dough is in pea-size pieces. In a separate bowl, whisk together the egg and heavy cream. Pour over the dry ingredients and mix with a fork until the dough begins to come together (it will still look quite shaggy). Dump out onto a lightly floured surface and knead the dough together. Wrap in plastic wrap and chill in the refrigerator for 30 minutes.

Meanwhile, to make the filling: Heat the oil in a large skillet over medium heat. Sauté the onion and jalapeño for 4 to 5 minutes, until the onion becomes fragrant. Add the zucchini and corn and continue to cook until both are beginning to brown, 6 to 8 minutes.

Combine the vegetables with the cilantro, salt, red pepper, and lime juice in a food processor. Pulse 4 or 5 times, bringing the mixture together but still leaving texture. Place in the refrigerator to let cool while you roll out the empanada dough.

Preheat the oven to 375°F (190°C, or gas mark 5). Line a baking sheet with parchment paper.

To assemble the empanadas: Whisk together the egg and water. Divide the dough into 12 equal pieces. Roll each piece into a 6-inch (15-cm) circle about ⅛ inch (3 mm) thick. Place ¼ to ⅓ cup (52 g) filling inside a circle, brush the edges with the egg wash, fold over the dough, and crimp the edges together. Place on the baking sheet, brush with more egg wash, and sprinkle with paprika. Repeat with the remaining dough pieces.

Bake for 25 minutes, or until the empanadas are golden brown. Let cool slightly before serving. Store in an airtight container in the refrigerator for 1 to 2 days.

WHEAT BERRIES

Before I started focusing on whole grains, I used a lot of all-purpose flour, not ever questioning what it was or why it was best for what I was making. It took me years to connect all-purpose flour to wheat flour to wheat berries. A connection that seems obvious now required quite a few steps. While the all-purpose flour found in stores does come from wheat, it has had all the nutrients stripped away and has often been bleached to remove any color of wheat. In other words, all-purpose flour is filler—and not a good one.

Wheat berries consist of three main layers: bran, germ, and endosperm. The majority of nutrients in the wheat berry live in the outer layer, the bran. In milling white flour, this bran layer is the first to be removed and is often sold separately. The grain then goes through a second milling to remove the germ, which also contains nutrients and is sold as a separate product. The end result is the endosperm, which is then milled and often bleached. Milling wheat berries at home allows you to keep all the nutrients in the flour.

As I started adding more whole wheat flours into my rotation, I researched the best ways to use them, and it became a bit disconcerting to see that every recipe still had half all-purpose flour to keep the texture the same. Every time I made something, I started pushing the limit, and I eventually ditched all-purpose flour. Sure, the items I make aren't exactly like their all-purpose flour counterparts, but they are delicious in their own way. I've fallen in love with using wheat flour; it adds a layer of heartiness and nutrition not found in all-purpose flour.

It may seem that with all the choices of wheat varieties, modern wheat berries would be the last on my list to use. However, I've found that wheat berries are the easiest to find in stores and make for the easiest transition from all-purpose flour. I keep varieties of wheat berries on my shelf, as each has found a place in different recipes.

Wheat berries are the third largest cereal crop, after rice and corn. These berries are thought to be one of the first domesticated cereal crops, tracing back thousands of years to Turkey. The wheat that is grown today is a hybridization tracing back to ancient emmer wheat and comes in a variety of forms.

Hard wheat berries and soft wheat berries have a major difference that can have a profound impact on baking. Hard wheat berries have a high protein content that makes the flour better suited for baking breads. Soft wheat berries are often ground for pastry flour and make wonderful flaky crusts, crumbly pastries, and airy cakes. The combination of the two is what makes up all-purpose flour.

To make the berries even more diverse, they are planted and harvested at different times of the year, and they also come in red or white. Try buying a few different types and experimenting to find your favorite flavor. Wheat berries can be found in the bulk bins in health food stores and even in some regular grocery stores.

Hard Wheat Flour

I remember a time when my mother was trying her best to switch the family over to whole wheat products. I was a big fan of my white bread and often referred to anything with whole wheat as "scary food." Since then, I have profusely apologized to my mother. As I matured in my ways of eating, I fell in love with the nutty, slightly earthy flavor that whole wheat flours add to breads, pasta, and pizza crust.

When I grind my own wheat, especially hard wheat, I tend to be a bit picky with my flavors. Although I love red hard wheat, I tend to cook and bake for people who still aren't crazy about the flavor and texture that comes with the red wheat. Most of the time, I grind white hard wheat. I find that the flavor is a bit milder and creates a slightly lighter texture in baked goods. I occasionally mix the two together in bread baking. White wheat has had the genes for the bran color removed to lose the reddish color. The removal of this coloring has no effect on nutrients but does change the flavor profile slightly.

Weights and Measurements

- 1 cup hard wheat berries = 180 g
- 1 cup hard wheat flour = 120 g
- 1 cup (180 g) hard wheat berries = 1½ cups (180 g) hard wheat flour

BASIC
HONEY WHEAT BREAD

I love having fresh bread around, but it doesn't last as long
as store-bought bread because of the lack of preservatives. However, stale bread makes
wonderful croutons: Cut the bread into ½-inch (1.2-cm) cubes, toss with olive oil
and a few herbs, and bake at 425°F (220°C, or gas mark 7) for 15 to 20 minutes.

YIELD: 12 TO 16 SLICES

1½ cups (355 ml) warm water (105°F to
110°F, or 40°C to 43°C)

1 package (2¼ teaspoons) active dry yeast

¼ cup (85 g) honey

4 to 4½ cups (480 to 540 g) white wheat
flour

¼ cup (60 ml) olive oil or melted butter,
plus more for brushing

1½ teaspoons sea salt

In the bowl of a stand mixer fitted with the dough hook attachment, combine the warm water, yeast, and honey. Let sit for 3 to 4 minutes, until the yeast becomes active and frothy. Stir in 2 cups (240 g) of the wheat flour, the olive oil, and salt. Start the mixer on low speed.

Continue to add flour, ¼ cup (30 g) at a time, letting the flour incorporate after each addition, until the dough pulls away from the side of the bowl. Continue to knead on low speed for 2 to 3 minutes after the dough pulls away. This will give you time to adjust the flour and let the dough knead. The dough should be soft and slightly sticky. Remove the dough hook, cover the mixing bowl with a damp towel, and set aside to rise for 1 to 1½ hours.

Once the first rise is over, remove the dough from the bowl and pat out into an 8 × 10-inch (20 × 25.5-cm) rectangle. Roll into a tight log and place seam side down in a 4½ × 8½-inch (11 × 21-cm) oiled bread pan. Cover again and set aside for about 1 hour. Preheat the oven to 375°F (190°C, or gas mark 5).

Once the loaf has risen the second time, brush with oil and bake for 35 to 40 minutes, until the bread has a golden crust and sounds hollow when tapped. Let cool in the pan for 10 minutes, then transfer to a wire rack to finish cooling before slicing. Store in a sealed bag at room temperature for 2 to 3 days, or slice and store in the freezer, thawing as needed.

For a variation, before rolling the dough into the log, brush on ¼ cup (55 g) melted butter and sprinkle with 2 tablespoons (14 g) cinnamon. This creates a beautiful cinnamon swirl loaf, perfect for morning toast or french toast.

CINNAMON
PULL-APARTS

I have such a weakness for bread and cinnamon together. I'll often
make my Basic Honey Wheat Bread (page 60) as a cinnamon swirl loaf to satisfy this craving, but sometimes I need more. These cinnamon pull-apart rolls are a wonderful weekend
breakfast to enjoy while snuggled up with a cup of hot coffee and a good book.

YIELD: 6 LARGE PULL-APARTS

For the dough:

½ cup (120 ml) warm whole milk (105°F to
110°F, or 40°C to 43°C)

2 tablespoons (30 ml) maple syrup

2½ teaspoons active dry yeast

2¼ to 2½ cups (300 g) whole wheat flour,
divided

2 large eggs

¼ cup (56 g) butter, melted

½ teaspoon sea salt

For the filling:

⅓ cup (76 g) butter, melted

¼ cup (60 g) packed brown sugar

¼ cup (50 g) organic cane sugar

2 tablespoons (14 g) ground cinnamon

To make the dough: In the bowl of stand mixer fitted with the dough
hook, combine the milk, maple syrup, and yeast. Stir, then let sit until the
yeast becomes active and frothy, 5 to 10 minutes. Add 1½ cups (180 g)
of the flour, the eggs, butter, and salt. Stir to combine.

Start the mixer on low speed and add ¼ cup (30 g) of additional flour
at time, letting it incorporate after each addition. Continue to add flour
until the dough begins to pull away from the side of the bowl. The dough
should be soft but not sticky. Remove the dough from the mixer, shape
into a ball, put back into the bowl, and cover with a damp towel. Place
the bowl in a warm, draft-free spot and let it rise until it is doubled in
size, about 1 hour.

To make the filling and assemble: Remove the dough from the bowl and
place onto a floured surface. Roll out to a 10 × 14-inch (24.5 × 35.5-cm)
rectangle and smear with the melted butter. Sprinkle on the sugars and
cinnamon. Cut the dough into 1 × 1-inch (2.5 × 2.5-cm) squares. Stack
4 or 5 squares on top of each other and place sideways in the cups of a
jumbo muffin tin, with the cut side of the squares facing out like a fan.
Cover and let rise for 45 minutes.

Preheat the oven to 375°F (190°C, or gas mark 5).

Bake the pull-aparts for 11 to 13 minutes, until soft. Serve warm. Store
in an airtight container at room temperature, or freeze for later use. To
reheat, preheat the oven to 350°F and brush the rolls with melted butter.
Heat for 4 to 6 minutes, until warm.

To make these in a standard-size muffin tin, divide the dough into 12 pieces
instead of 6 and follow the remaining steps. Reduce the baking time to 8 to
10 minutes.

GRILLED VEGETABLE TACOS WITH WHOLE WHEAT TORTILLAS

Tacos are one of my favorite meals, and occasionally I like to switch up my tortillas. These whole wheat tortillas are hearty and make for a great taco base. For a variation, try adding a pinch of ground cumin and a squeeze of lime juice while mixing the tortilla ingredients.

YIELD: 6 TACOS

For the tortillas:

1½ cups (180 g) whole wheat flour

½ teaspoon sea salt

¼ teaspoon baking powder

¼ cup (60 ml) olive oil

¾ cup (175 ml) warm water (105°F to 110°F, or 40°C to 43°C)

For the filling:

1 medium (200 g) zucchini

1 medium (150 g) red onion

1 tablespoon (15 ml) olive oil

1 jalapeño chile pepper

Juice of 1 lime

1 tablespoon (20 g) honey

¼ cup (4 g) chopped fresh cilantro, plus more for garnish

½ cup (90 g) cooked black beans, rinsed and drained if using canned

2 cups (110 g) chopped lettuce, such as romaine or green leaf

2 ounces (55 g) queso fresco, crumbled

To make the tortillas: In a large bowl, combine the flour, salt, and baking powder and stir to combine. Add the oil and, using your fingers, rub it into the flour mixture. Pour in the warm water and mix until a ball forms, using your hands as needed. Cover and let rest for 20 minutes.

Preheat a griddle or large skillet over medium heat. Divide the dough into 6 balls and, on a floured surface, roll each ball into a circle about 6 inches (15 cm) in diameter and ⅛ inch (3 mm) thick. (Alternatively, line a tortilla press with waxed or parchment paper, place a dough ball in the center, and press the plates together.) Cook 1 tortilla at a time on each side for 2 to 3 minutes, or until the tortilla begins to blister and brown on each side. Continue with the remaining dough.

Stack the tortillas in a damp towel until ready to use.

To make the filling: Preheat a grill or grill pan. Slice the zucchini into strips ¼ inch (6 mm) thick and 2 inches (5 cm) long and the onions into ¼-inch (6-mm) circles. Brush on the olive oil and grill the zucchini, onion, and jalapeño on each side until blistering, 3 to 5 minutes depending on the grill heat. Dice the onion and jalapeño. (Remove the chile seeds for less heat.)

In a small bowl, whisk together the lime juice, honey, and cilantro. Toss the dressing with the grilled vegetables and black beans.

To assemble the tacos, divide the grilled vegetables among the tortillas, then top with the lettuce, crumbled cheese, and extra cilantro if desired. Fold in half or roll up. Store leftover tortillas in a sealed bag in the refrigerator or freezer.

🖊 If using the tortillas after they have cooled, wrap them in a damp kitchen towel and place in a warm oven until heated through.

SOFT WHEAT FLOUR

I am particular about the type of soft wheat I use. I use soft wheat to grind pastry flour for use in everything except breads. Soft wheat pastry flour makes for perfect pancakes, scones, and piecrust. The difference between winter and spring wheat is minimal for me, but I have better luck with white over red. Grinding soft white wheat also makes for an easier transition if you're baking for someone who doesn't necessarily like the flavor of wheat.

I also receive quite a few questions about subbing whole wheat flour for all-purpose flour in recipes. The general rule is to only sub half so as not to affect texture and taste. However, I've been subbing 100 percent whole wheat flour, especially soft wheat flour, and found that although it does somewhat affect texture and taste, I haven't missed the all-purpose flour.

Look for soft wheat berries in the bulk bins at health food stores. Occasionally the berries won't be marked to identify the color or season, but many times a check by a store employee will give you the answer.

Weights and Measurements

1 cup soft wheat berries = 180 g

1 cup whole wheat flour = 120 g

1 cup (180 g) soft wheat berries = 1½ cups (180 g) whole wheat flour

THREE-ONION AND BLUE CHEESE TURNOVERS

There was a large part of my life when I associated pie dough only with sweet desserts or breakfast pastries. However, when I first learned to make quiche, it opened the doors wide to what I could do with a basic piecrust. These turnovers are one of my favorite savory ways to use pie dough.

YIELD: 6 TURNOVERS

For the dough:

1½ cups (180 g) whole-wheat pastry flour

2 tablespoons (26 g) organic cane sugar

¼ teaspoon sea salt

½ cup (112 g) cold butter, cubed

2 tablespoons (30 ml) cold water

For the filling:

2 tablespoons (30 ml) olive oil

1 large (200 g) red onion, sliced into ¼-inch (6-cm) slices

2 shallots (70 g), thinly sliced

5 or 6 scallions (60 g), diced

1 tablespoon (15 ml) balsamic vinegar

2 ounces (55 g) blue cheese, crumbled

To make the dough: In a food processor or large bowl, combine the pastry flour, sugar, and salt. Pulse or whisk to combine. Cut in the butter, either pulsing or using a pastry blender, 2 knives, or your hands, until the dough is in pea-size pieces. A tablespoon (15 ml) at a time, add the cold water and pulse or stir with a fork until the dough comes together. Add a bit more cold water, as needed, to bring the dough together. Remove from the bowl, pat the dough into a disk, wrap in plastic wrap, and refrigerate for 20 minutes.

Meanwhile, to make the filling: In a medium skillet over medium-low heat, heat the olive oil. Add the onion and shallots and cook for 10 minutes, stirring occasionally, turning the heat down to low if the onions start to brown quickly. Add the scallions and cook for 5 minutes more. Stir in the balsamic vinegar and cook for 1 minute. Remove from the heat and transfer to a bowl. Place in the refrigerator and let cool completely. When ready to use, mix in the blue cheese.

Preheat the oven to 375°F (190°C, or gas mark 5). Lightly grease a baking sheet, or line with parchment paper.

On a floured surface, roll the dough into a 12 × 16-inch (30 × 40-cm) rectangle about ¼ inch (6 mm) thick. Cut into 6 squares. Divide the filling evenly and place in the center of each square. Carefully fold the dough over into a triangle and crimp the edges with a fork.

Transfer the turnovers to the prepared baking sheet and bake for 18 to 22 minutes, or until browned and crispy. Let cool slightly before serving.

To make scrumptious appetizers or finger foods, cut the dough into 12 squares to make mini turnovers. Reduce the baking time to 15 to 18 minutes.

FRIED EGG AND GARLICKY GREENS BISCUIT SANDWICHES

One of my biggest weaknesses is biscuits, especially when
it comes to an indulgent breakfast sandwich. While I don't eat nearly as many
biscuits as I used to, these sandwiches hit the spot when I'm craving a delicious and
decadent breakfast. These sandwiches are also perfect for an on-the-go breakfast
(just cook the egg until the yolk is firm).

YIELD: 4 SANDWICHES

For the biscuits:

1¼ cups (150 g) whole-wheat pastry flour

1½ teaspoons baking powder

¼ teaspoon sea salt

¼ cup (56 g) cold butter, cut into pieces

½ cup (50 g) finely chopped scallions

¼ cup (8 g) grated Parmesan, plus more
  for topping

1 large egg

2 tablespoons (30 ml) buttermilk

1 tablespoon (20 g) honey

2 tablespoons (30 ml) heavy cream

For the filling:

1 tablespoon (15 ml) olive oil

1 clove garlic, minced

2 cups (110 g) loosely packed greens,
  coarsely chopped (see Note)

1 tablespoon (14 g) butter

4 large eggs

3 to 4 tablespoons (33 to 44 g) spicy
  brown mustard

Preheat the oven to 425°F (220°C, or gas mark 7).

To make the biscuits: In a large bowl, combine the pastry flour, baking powder, and salt; stir to combine. Cut in the butter using a pastry blender, 2 knives, your hands, or a food processor until the dough is in pea-size pieces. Stir in the scallions and cheese.

In a smaller bowl, whisk together the egg, buttermilk, and honey. Combine with the dry ingredients until the dough pulls together. Transfer from the bowl to a floured surface. Pat the dough into a ½-inch (1.2-cm) thickness and cut into desired shapes (squares or circles if using a biscuit cutter). Transfer the biscuits to a baking sheet. Brush the biscuits with the cream and sprinkle with more Parmesan. Bake for 14 to 18 minutes, until golden.

Meanwhile, to make the filling: Heat the olive oil in a large skillet over medium-high heat. Add the garlic and cook for 1 minute. Add the greens, toss a few times, cover, and remove from the heat. Let sit for 5 minutes, or until the greens have wilted slightly.

Remove the greens from the skillet. Return the skillet to medium-high heat and add the butter. Fry the eggs, flipping once, until they reach the desired doneness. Assemble the sandwich by cutting a biscuit in half and layering on 1 egg, one-quarter of the greens, and about 1 teaspoon spicy brown mustard.

You can use any type of greens for this recipe, but I've found that spinach or Swiss chard works best.

RHUBARB PIE
WITH WHOLE WHEAT CRUST

In my family, rhubarb is king (or queen). During the late
spring months, we used to watch as the rhubarb sprouted from the ground,
and we counted the days until we could use it in pie or rhubarb sauce. We've even gone
so far as to transplant a rhubarb plant that was first planted by my great-grandfather
everywhere we've moved. I love the contrast between the sweet and the tart
in this rhubarb pie and the earthy whole wheat crust.

YIELD: 6 TO 8 SERVINGS

For the crust:

1½ cups (180 g) soft whole wheat flour

2 tablespoons (26 g) organic cane sugar

¼ teaspoon sea salt

½ cup (112 g) cold butter, cubed

2 to 3 tablespoons (28 to 45 ml) cold
    water

For the filling:

1½ pounds (680 g) rhubarb, cut into
    ½-inch (1.2-cm) dice (6 cups)

½ cup (100 g) organic cane sugar

½ cup (115 g) packed brown sugar

½ cup (60 g) soft wheat flour

1 tablespoon (15 ml) fresh lemon juice

1 teaspoon ground cinnamon

1 tablespoon (14 g) butter, melted

2 tablespoons (26 g) organic cane sugar

Preheat oven to 375°F (190°C, or gas mark 5).

To make the crust: Combine the whole wheat flour, sugar, and salt in a
medium bowl and whisk to combine. Cut in the cubed butter with your
hands or 2 knives until the dough is in pea-size pieces. Using a fork, stir
in 2 tablespoons (28 ml) cold water, and add about 1 more tablespoon
(15 ml), or as needed to bring the dough together—but be cautious of
making the dough too wet. Transfer from the bowl to a floured surface
and divide into 2 balls. Form each into a disk, cover with plastic wrap,
and refrigerate for 20 minutes.

To make the filling: In a large bowl, combine the rhubarb with the
sugars, soft wheat flour, lemon juice, and cinnamon. Toss together
until the rhubarb is coated. Set aside while the dough chills.

On a lightly floured surface, roll out each dough ball into an 11- to
12-inch (28- to 30-cm) circle. Drape 1 round of dough into a 9-inch
(23-cm) pie pan, being careful not to stretch or tear it. Pour in the
rhubarb filling, and top with the remaining crust. Crimp the edges
with a fork or your hands. Cut an × into the top of the pie, brush on the
melted butter, and sprinkle the sugar on top. Bake for 50 to 60 minutes,
until the crust is golden and the filling is bubbly. Let cool in the pan on
a wire rack before slicing. Store the pie refrigerated in an airtight
container or covered with plastic wrap for 4 to 5 days.

DURUM WHEAT
BERRIES

Nearly all the grains, legumes, and nuts in my house serve a dual purpose: as both a whole food and as flour. However, I keep durum wheat berries on hand for two very specific purposes: pasta and pizza. Nearly every store-bought packaged pasta lists durum flour as the main ingredient, either as itself or as semolina.

Durum, from the Latin for *hard*, is the hardest of all the wheat berries and has the highest protein content. However, even with the high protein levels, its gluten levels are low. Durum wheat was created by selection from domesticated emmer wheat. The durum wheat berry has a smooth, nutty taste similar to emmer and is slightly larger than modern wheat. Durum berries are widely cultivated for their use in commercial pastas; however, finding the berries for personal use can be a bit tough. Few health food stores carry durum, but a quick Internet search will yield a few places to order from online.

Durum Flour

Home-ground durum flour's texture may be a bit confusing at first. Instead of having a soft, somewhat silky texture like other ground wheat, durum flour is coarse, resembling cornmeal. Semolina flour can also be confusing, as it is ground from durum wheat. This flour is not considered a whole-grain flour, as it is ground from only the endosperm of the durum wheat berry. Grinding durum at home guarantees flour that is 100 percent whole grain.

Because of its low gluten levels, durum flour is best used for pastas and products not containing yeast. Durum flour can be used in breads, but using 100 percent durum flour will result in a denser end product. (My only exception to this guideline is pizza dough, as I love the hearty, slightly chewy pizza crust that comes from using all durum flour.) Pasta made from durum flour cooks up tender, with a delightful and lighter wheat flavor compared with other types of wheat.

Durum flour is best ground in a grain mill or high-speed blender. If you sift durum flour after grinding it, you may notice the bran remains in the sieve. I experimented with sifting some of the bran out of the flour before using it, but the difference wasn't noticeable enough for me to continue sifting out the bran.

Weights and Measurements

1 cup durum wheat berries = 180 g

1 cup durum flour = 120 g

1 cup (180 g) durum wheat berries = 1½ cups (180 g) durum flour

FIG AND BLUE CHEESE PIZZA

Although I don't often use 100 percent durum flour for
yeasted baked goods, the flavor is too good to pass up when it comes to
pizza dough. This crust is hearty and dense, perfect for flavorful ingredient
combinations such as the fig and blue cheese here. For a bit of extra crunch,
replace ½ cup (60 g) of the durum flour with ½ cup (70 g) cornmeal.

YIELD: 3 OR 4 SERVINGS

For the crust:

¾ cup (175 ml) warm water (105°F to
110°F, or 40°C to 43°C)

2 tablespoons (26 g) organic cane sugar

1 package (2¼ teaspoons) active dry yeast

2½ cups (300 g) durum flour, divided

2 tablespoons (30 ml) olive oil

½ teaspoon sea salt

For the topping:

2 tablespoons (30 ml) olive oil

2 cloves garlic, minced

10 to 12 fresh figs (150 to 180 g),
sliced ¼ inch (6 mm) thick

4 ounces (115 g) Gorgonzola, crumbled

2 tablespoons (15 g) walnuts, toasted and
chopped

1 to 2 tablespoons (20 to 40 g) honey

To make the crust: In a large bowl, combine the water, sugar, and yeast
and stir with a wooden spoon until the yeast is dissolved. Let sit until
the yeast begins to activate and froth, 5 to 10 minutes. Once the yeast
is ready, add 2 cups (240 g) of the flour, the olive oil, and salt. Stir the
dough until it comes together, adding more flour 2 to 4 tablespoons
(15 to 30 g) at a time, mixing the dough for a bit until the flour is
incorporated. Once the dough is barely sticky, cover with a damp cloth
and let rise for 1½ hours.

Once the dough has risen, preheat the oven to 450°F (230°C, or gas
mark 8).

To make the topping and assemble: Roll and stretch the dough into a
12- to 14-inch (30- to 35.5-cm) circle on a pizza pan. Brush with the olive
oil and sprinkle with the garlic. Layer the figs around the pizza. Sprinkle
with the Gorgonzola. Bake for 12 to
15 minutes, until the crust is crisp and the cheese has melted.

Serve the pizza with sprinkled with the walnuts and drizzled with
the honey.

Black Pepper Pasta with Goat Cheese and Pesto

When I've had a stressful day or just an overall grumpy one,
I make pasta. The motions of whisking the flour into the egg, kneading the dough,
and cutting it into strips or shapes can lift any ill mood I may be feeling at the time;
it's very therapeutic. Black pepper pasta is a favorite of mine. The extra bite from the
fresh pepper gives a bit more life to the pasta and pairs well with many
different toppings. Fresh herbs also work well. Try mixing chopped
fresh rosemary, basil, or oregano into the dough.

YIELD: 4 SERVINGS

For the pasta:

2 cups (240 g) durum flour

1 tablespoon (5 g) cracked black pepper

½ teaspoon sea salt

2 large eggs

2 tablespoons (30 ml) water

1 tablespoon (15 ml) olive oil

For the pesto:

2 cloves garlic, peeled

¼ cup (35 g) pine nuts

⅓ cup (10 g) grated Parmesan

2 cups packed (80 g) fresh basil leaves

3 tablespoons (45 ml) olive oil

2 tablespoons (30 ml) fresh lemon juice

2 ounces (55 g) goat cheese, crumbled

To make the pasta: Combine the durum flour, pepper, and salt on a clean, flat surface. Make a well in the middle and add the eggs and water to the center. Using a fork, whisk the eggs with the water and slowly begin to incorporate the flour. Continue to combine the flour and eggs until a paste-like texture forms. Keep mixing, eventually trading the fork for your hands, and knead the dough into a smooth ball. Cover with a damp towel and let rest for 20 minutes.

Divide the dough into 8 pieces. On a floured surface, or using the pasta attachment of a stand mixer, roll out each piece to a ¹⁄₁₆-inch (1.5-mm) thickness. Using a knife or pizza cutter, cut pasta strips that are ¼ inch (6 mm) wide.

Bring a large pot of salted water to a boil and add the noodles. Cook until the noodles float to the top and are tender, 4 to 5 minutes. Drain and toss with the olive oil.

To make the pesto: Add the garlic to a food processor and pulse until minced. Add the pine nuts and pulse 2 or 3 times to incorporate into the garlic. Next, add the Parmesan, basil, olive oil, and lemon juice. Run the food processor until a pesto forms. Taste and add more olive oil or lemon juice as desired.

Toss the warm pasta with the pesto and serve with the goat cheese sprinkled over the top.

Spaghetti with Tomato Basil Sauce

Homemade pasta was one of the first things to make
a dent in my unprocessed kitchen. As a holiday gift, my father gave me the pasta
attachment for my KitchenAid mixer, and I still tell him, to this day, that it's the best gift
he's ever given me. I use it at least once a week. It turns out that homemade pasta and
tomato sauce are so much better than the boxed pasta and jarred spaghetti sauce
I remember as a child. Don't fret if you don't have a pasta roller; simply use the same
technique that is described in the recipe for Black Pepper Pasta on page 73.

YIELD: 4 SERVINGS

For the sauce:

2 pounds (910 g) Roma tomatoes, halved

1 medium (160 g) onion, cut into large chunks

2 cloves garlic, finely chopped

3 tablespoons (45 ml) olive oil, divided

¼ cup (10 g) fresh basil leaves

2 tablespoons (30 ml) fresh lemon juice

1 tablespoon (20 g) honey

For the pasta:

2 cups (240 g) durum flour

½ teaspoon sea salt

2 large eggs

2 tablespoons (30 ml) water

1 tablespoon (15 ml) olive oil

¼ cup (8 g) grated Parmesan (optional)

Preheat the oven to 400°F (200°C, or gas mark 6).

To make the sauce: In a medium bowl, toss together the tomatoes, onion, and garlic with 1 tablespoon (15 ml) of the olive oil. Spread out in a single layer in a roasting pan. Roast for 30 to 40 minutes, until the tomatoes are tender and starting to char. Remove from the oven and let cool slightly. Place the tomato mixture in a food processor and add the basil, the remaining olive oil, the lemon juice, and honey. Pulse the mixture until a sauce forms.

To make the pasta: Combine the durum flour and salt on a clean, flat surface. Make a well in the middle and add the eggs and water to the center. Using a fork, whisk the eggs with the water and slowly begin to incorporate the flour. Keep mixing, eventually trading the fork for your hands, and knead the dough into a smooth ball. Cover with a damp towel and let rest for 20 minutes.

Divide the dough into 8 pieces. Working with 1 piece at a time, flatten the dough with your hands, making sure it is covered with flour. Using an electric or hand-crank pasta roller, roll the dough into thin sheets. Replace the flat roller with the spaghetti cutter and run the dough through again to cut it.

Bring a large pot of salted water to a boil and add the spaghetti. Cook until the spaghetti floats to the top and is tender, 4 to 5 minutes. Drain and toss with the olive oil. Serve the pasta with 1 or 2 ladles of sauce and a sprinkling of Parmesan if desired.

CHAPTER 3

Gluten-Free Grains

The grains and seeds in this chapter have risen in popularity as the buzzword "gluten-free" has nosed its way into our everyday lives. Although these grains are indeed good for those who are required to follow a gluten-free diet, they can also be a wonderful addition to a regular diet. Each grain has a unique flavor and texture that makes it fun to cook with and, of course, grind into flour.

Because these grains lack the proteins of glutenous wheat, they don't always act like you would expect. As such, using gluten-free flours can be a bit challenging at times, but there are little tricks you can employ to make working with them easier. I've sprinkled these tips throughout the chapter, as each grain has its own characteristics, but here are a few overall things to keep in mind when working with gluten-free flours:

- Gluten-free baked goods often don't rise the way gluten-containing goods do.

- The texture of gluten-free goods can be dense or crumbly, because gluten-free flours lack the stretch and elasticity that gluten yields.

- Doughs do not always want to hold together through the assembly process. Be gentle and patient when working with these flours. Employing a little extra care will go a long way toward getting a good result.

- These doughs tend to be more sticky or crumbly. When rolling them out, use enough flour, because the dough tends to tear and fall apart more than when working with gluten-based flours.

- Each grain's flavor will come through in the flour and can affect the overall dish.

- Go by the weight of the flours rather than depend on the volume measurements. Different flours have different weights per cup, and this could easily throw off a recipe if you try to substitute in a different flour.

- Often, a mix of 70 percent flour blends and 30 percent starch blends are useful to achieve the lightness of traditional baked goods. I've created a few recipes with and without starches based on the textures I like best. Also, these recipes can be adapted to use a blend of gluten-free flours if the taste of one isn't to your liking. My favorite combination is oat, buckwheat, and millet, but feel free to experiment and find your own.

TEFF

I wasn't that familiar with teff until I started learning more about gluten-free baking. I noticed the flour showing up in gluten-free flour blends, which piqued my curiosity. Ever since, teff flour has been my go-to addition for baked goods, especially chocolate-based goodies. The darker teff seeds grind into a flour that has a rich flavor similar to molasses, while the lighter seeds have a nuttier flavor, perfect for pairing with walnuts, pecans, and hazelnuts. I also keep teff around my kitchen for use in breakfast porridges, as the seed cooks up into a creamy, delicious base for additional ingredients.

Thought to have originated in Ethiopia, teff is now grown and cultivated around the world. Teff isn't actually a true grain but is a pseudo-grain, like millet and quinoa. It's a grass seed, similar in size to a poppy seed, ranging in color from a dark reddish brown to ivory. Most teff I've purchased in health food stores is dark, which makes for a dark brown flour. Of all the grains, teff has the highest amount of calcium, and it contains vitamin C, protein, and iron. Teff makes for a great crop not only because of its nutrients but also because of its ability to thrive in poor soil, droughts, and floods.

Finding teff in seed form can be a bit challenging. Most health food stores carry teff flour, but finding the seeds is hit or miss. However, many online stores carry teff seeds and can easily ship to your door.

Teff Flour

Teff flour has become a staple in many of my gluten-free recipes, even though it is a relatively new addition to my kitchen. The taste of the grain carries over to the flour, which can add an extra touch of richness that makes baked goods a bit more special. When I first started experimenting with teff flour, I found many cake and brownie recipes that called for it because its molasses flavor complements chocolate. I've also found that teff flour pairs well with earthy vegetables such as beets and carrots.

In general, I try to grind only the amount of flour I need for a recipe in order to keep as many nutrients locked in as possible. I find this especially true for teff, not only because of the nutrients but also because of the flavor. I use my coffee grinder or a high-speed blender to grind small batches of teff. You may use a grain mill, but be cautious. The tiny seeds have the ability to clog the grinder and bring the mill to a stop. For all methods of grinding, I recommend sifting the flour through a sieve to catch any small seeds that did not grind.

Weights and Measurements

1 cup teff grain = 180 g

1 cup teff flour = 152 g

1 cup (180 g) teff grain = about 1 cup + 3 tablespoons (180 g) teff flour

CURRIED SWEET POTATO
AND TEFF BURGERS

One of my summer rituals is to make batches of 2 or 3
different kinds of veggie burgers and freeze them for all the future times we grill.
I love the flavor of curry, and the teff flour pairs perfectly with it. I recommend
serving this burger with smashed avocado and cilantro pesto.

YIELD: 4 TO 6 BURGERS

1 medium (250 g) sweet potato, peeled

¼ cup (38 g) teff flour

2 tablespoons (13 g) curry powder

½ teaspoon sea salt

½ teaspoon freshly ground black pepper

2 tablespoons (30 ml) fresh lime juice

1 can (15 ounces, 1½ cups, or 240 g)
chickpeas, rinsed and drained

¼ cup (25 g) chopped pecans

Bring a pot of water to a boil. Cut the sweet potato into 1-inch
(2.5-cm) cubes, add to the boiling water, and cook until tender,
10 to 15 minutes.

Preheat the oven to 375°F (190°C, or gas mark 5).

In a food processor, combine the cooked sweet potato, teff flour, curry
powder, salt, pepper, and lime juice. Pulse until combined and smooth.
Add the chickpeas and pecans, pulsing a few more times until combined.

Remove the mixture from the food processor. Run your hands under
water, and shape the mixture into 4 to 6 patties, depending on desired
size. If grilling, bake in the oven for 15 minutes in order to get a head
start on the burger. For grilling, place the pre-baked burgers on a hot grill
and cook for 3 to 5 minutes on each side, depending on the heat of the
grill, until the burgers are crisp and browned.

If panfrying, there is no need to bake first. Place the burgers in a lightly
oiled skillet over medium-low heat and cook for 6 to 8 minutes on each
side. The outside should be browned and crisp. Serve with your favorite
burger toppings.

If making ahead of time, follow the steps through forming the patties. Let cool,
and freeze the burgers, each separated by a square of parchment paper. Cook
directly from the freezer, without thawing.

RICOTTA AND BEET GALETTE WITH TEFF CRUST

With the sweetness of roasted beets and the heartiness of teff, this galette
is one of my favorites for flavor and for beautiful color! The dough and
the beets can be made up to 1 day ahead of time.

YIELD: 6 TO 8 SERVINGS

For the beets:

1 pound (455 g) mixed red and golden
   beets
1 tablespoon (15 ml) olive oil
⅛ teaspoon sea salt

For the crust:

½ cup (78 g) teff flour
¼ cup (36 g) sweet rice flour
¼ cup (32 g) cornstarch
¼ cup (32 g) arrowroot
1 tablespoon (13 g) organic cane sugar
½ teaspoon sea salt
½ cup (112 g) cold butter, cut into pieces
2 to 3 tablespoons (30 to 45 ml) water

For the filling:

1 cup (246 g) ricotta
¼ cup (40 g) crumbled feta
2 teaspoons (40 g) honey
2 tablespoons (3.4 g) minced fresh
   rosemary
¼ teaspoon sea salt
1 clove garlic, minced
1 tablespoon (15 ml) olive oil

1 egg yolk
1 tablespoon (15 ml) heavy cream or water

To make the beets: Preheat the oven to 400°F (200°C, or gas mark 6). Peel the beets and slice into ¼-inch (6-mm) slices. Toss with the olive oil and salt. Roast until tender, 40 to 50 minutes.

Meanwhile, to make the crust: In a food processor, combine the flours, cornstarch, arrowroot, sugar, and salt. Pulse in the butter until the dough is in slightly larger than pea-size pieces. Pulse in 1 tablespoon (15 ml) of the water, and continue to add water and pulse until dough starts to come together. Remove from the food processor and shape into a disk. Wrap in plastic wrap and refrigerate for 20 minutes.

To make the filling: In a medium bowl, mix together the ricotta, feta, honey, rosemary, salt, and garlic.

On a surface lightly covered with cornstarch, roll the dough out to a 12- to 14-inch (30- to 35.5-cm) circle. Carefully fold, transfer to a baking sheet lined with parchment paper, and pat back together any tears that may have formed. Spread the ricotta mixture evenly over the crust, leaving a 2-inch (5-cm) edge. Layer the beets evenly in a circular pattern, covering the ricotta. Fold the edges of the crust over the outer edge of the beets, pleating as needed to make an even circle. (The center of the galette won't have any dough over it.) Brush the olive oil over the beets.

Combine the egg yolk with the heavy cream or water and brush the outer edge of the crust. Bake for 40 to 45 minutes, until the crust is golden brown. Allow to cool slightly before serving.

🖊 If you don't own a food processor, simply combine all the dry ingredients in a bowl and cut in the butter with 2 knives or your hands until the dough is in pea-size pieces.

MINI CHOCOLATE BUNDT CAKES WITH PEANUT BUTTER FROSTING

I would feel remiss if I didn't provide one chocolate recipe
with teff flour. I have such a sweet tooth for chocolate and peanut butter together.
These cakes are wonderfully moist, and the peanut butter frosting puts them over the top.
Even though my family doesn't avoid gluten, we'll take a batch of these cakes
over a gluten-filled chocolate cake any day of the week.

YIELD: 6 MINI CAKES

For the cakes:

½ cup (76 g) teff flour

¼ cup plus 2 tablespoons (30 g) unsweetened cocoa powder

¼ cup (34 g) sorghum flour

2 tablespoons (12 g) oat flour

2 tablespoons (16 g) arrowroot

½ cup (115 g) packed brown sugar

¾ teaspoon baking soda

¼ teaspoon sea salt

¾ cup (168 g) whole-milk plain Greek yogurt

¼ cup (60 ml) 2% or whole milk

¼ cup (56 g) butter, melted

¼ cup (60 ml) maple syrup

2 large eggs

For the frosting:

1 cup (120 g) confectioners' sugar

¼ cup (65 g) natural creamy peanut butter

1 tablespoon (20 g) honey

1 teaspoon vanilla extract

2 to 3 tablespoons (30 to 45 ml) whole milk

Preheat the oven to 350°F (180°C, or gas mark 4). Lightly butter a jumbo 6-cup muffin pan.

To make the cakes: In a large bowl, stir together the teff flour, cocoa powder, sorghum flour, oat flour, arrowroot, brown sugar, baking soda, and salt. In a separate bowl, whisk together the yogurt, milk, butter, maple syrup, and eggs. Pour the wet ingredients into the dry ingredients and stir until any lumps are gone.

Divide the batter among the muffin cups and bake for 18 to 20 minutes. The tops of the cakes should spring back lightly when pressed. Let cool slightly, about 5 minutes. Run a knife along the outside edges and flip the muffin pan upside down onto a flat surface covered with waxed paper. Tap the bottom to pop the cakes out. Allow the cakes to cool before frosting.

To make the frosting: In a small bowl with an electric mixer, cream together the confectioners' sugar, peanut butter, honey, and vanilla. Add enough of the milk to reach a soft frosting consistency. Spoon the frosting on top of the cooled cakes and spread to cover them. Store the cakes in an airtight container at room temperature for 2 to 3 days.

While I prefer frosting, adding a bit more milk will create a thick icing that may be used when the cakes are still slightly warm.

BROWN RICE

Brown rice has always been a staple in my house. I keep it on hand for a quick rice and beans dinner or to whip up paella to feed a crowd. The mild flavor pairs so well with so many ingredients that I find myself constantly reaching for brown rice. Also, I have found that brown rice is a good way to ease people into eating whole grains.

Rice is the primary food eaten around the world. But for Westerners, it often comes with negative connotations when comparing its nutritional benefits with other grains. Brown rice does have nutritional value, primarily dietary fiber. It's considered a whole grain, as only the hull is removed. White rice, on the other hand, is a processed version of brown rice where the bran and part of the germ is polished away, leaving very few nutrients. Always choose any type of brown rice over white. I buy organic to try to limit my exposure to the arsenic that can potentially be found in rice, due to contaminated water and soil where rice is grown.

Rice comes in various sizes, colors, and flavors. I almost always keep short-grain and long-grain brown rice on hand. Short-grain brown rice contains more starch and therefore cooks up slightly sticky. I use it to make paellas and other creamy rice dishes. Long-grain makes perfect rice side dishes and salads. I also keep a jar full of sweet brown rice (page 128).

Brown rice is easily found in supermarkets and health food stores, both in bulk bins and prepackaged. I recommend purchasing the type of rice you will use most often and use the same kind to grind into flour.

Brown Rice Flour

When grinding brown rice flour, I primarily stick with short-grain brown rice. I have found that its extra starch content aids in the lightness of baked goods, especially when paired with other starches such as arrowroot. Long-grain brown rice flour also works well, but there may be a slight texture difference in the final product.

Though brown rice flour is a staple for many gluten-free baked goods, it's also useful in cooking. I keep extra brown rice flour on hand to thicken soups and roux or to mix in with other flours to make a wonderful breading. The light, nutty flavor pairs well with many foods and other flours.

I use my grain mill to grind both long- and short-grain brown rice; however, a high-speed blender or coffee grinder would work as well. I found my food processor didn't have enough power to break the grain down fine enough for flour. Brown rice flour can sometimes leave a gritty texture in baked goods, but grinding it fine enough helps to mitigate this issue. Also, mixing the brown rice flour with other gluten-free flours, such as oat or buckwheat, can help.

Weights and Measurements

1 cup brown rice = 180 g

1 cup brown rice flour = 144 g

1 cup (180 g) brown rice = 1¼ cups (180 g) brown rice flour

BROWN RICE CREPE MANICOTTI

One of the biggest challenges I hear from my friends who eat
gluten-free is finding good, gluten-free Italian food. By using brown rice in the crepes
for the manicotti, this becomes a delicious, gluten-free dish that everyone will enjoy
without the fuss of having to make homemade gluten-free pasta.

YIELD: 4 SERVINGS

For the tomato sauce:

1 tablespoon (15 ml) olive oil

2 cloves garlic, minced

3 large (400 g) tomatoes, diced

1 tablespoon (15 ml) balsamic vinegar

1 tablespoon (20 g) honey

¼ teaspoon sea salt

3 tablespoons (8 g) shredded fresh basil

For the crepes:

½ cup (72 g) brown rice flour

1 large egg

¼ cup plus 2 tablespoons (90 ml) whole milk

1 tablespoon (14 g) butter, melted

2 teaspoons honey

¼ teaspoon sea salt

For the filling:

1 cup (240 g) ricotta

1 cup (112 g) shredded mozzarella, divided

1 tablespoon (4 g) minced fresh oregano

1 clove garlic, minced

1 egg white

To make the sauce: In a large pot, heat the olive oil over medium-low heat. Add the garlic and cook for 1 minute. Add the tomatoes and cook until tender, 10 to 12 minutes. Stir in the vinegar, honey, salt, and basil. Cook for 2 more minutes. Remove from the heat and purée the sauce with an immersion or regular blender. Return the sauce to the stove and simmer over low heat while making the manicotti.

To make the crepes: Combine all the crepe ingredients in a blender until the mixture is smooth. Heat a lightly buttered 8-inch (20-cm) skillet over medium-low heat. Spoon about 3 tablespoons (45 ml) of the mixture into the skillet and move the pan in a circular motion until the batter covers the pan. Cook for 1 to 2 minutes, until set. Flip and cook for 30 seconds more. Transfer to a plate and repeat with remaining the batter, overlapping the finished crepes on the plate.

Preheat the oven to 375°F (190°C, or gas mark 5).

Spoon one-third of the tomato sauce into an 8 × 8-inch (20 × 20-cm) or 9 × 9-inch (23 × 23-cm) baking pan so the sauce covers the bottom. Stir together the ricotta, ½ cup (56 g) of the mozzarella, the oregano, garlic, and egg white. Set 1 crepe on a work surface. Spoon 2 to 3 tablespoons (97 g) of the ricotta mixture into the center of the crepe, roll the crepe edges over the filling, and place seam side down in the pan. Repeat with remaining crepes and filling. Evenly pour the remaining tomato sauce over the crepes and sprinkle with the remaining ½ cup (56 g) mozzarella. Bake the manicotti for 30 to 40 minutes, until the mozzarella has melted and is golden.

When making the crepes, move quickly when pouring the batter and swirling the pan. It is easy to cook the underside of the crepe before finishing the swirl, resulting in an incomplete crepe.

PEANUT BUTTER COOKIES WITH BROWN RICE FLOUR

In recent years, I have become what one would call a peanut butter fanatic. I eat it most days for breakfast, I smear it on bananas for snacks, and my favorite ice cream has a large swirl of peanut butter in the center. These peanut butter cookies are similar to the ones I've always eaten, with a few updates to make them gluten-free.

YIELD: 12 LARGE COOKIES

½ cup (112 g) butter, softened

½ cup (100 g) organic cane sugar

½ cup (115 g) packed brown sugar, preferably dark brown

⅓ cup (88 g) natural creamy peanut butter

1 large egg

½ teaspoon vanilla extract

½ teaspoon baking soda

⅛ teaspoon sea salt

¾ cup (108 g) brown rice flour

¼ cup plus 2 tablespoons (52 g) sorghum flour

¼ cup plus 2 tablespoons (38 g) oat flour

¼ cup (30 g) tapioca starch

¼ cup (38 g) sweet brown rice flour, plus extra, divided

Preheat the oven to 375°F (190°C, or gas mark 5). Lightly grease a baking sheet (or 2) or line it with parchment paper.

In a large bowl or the bowl of a stand mixer, cream together the butter and sugars. Add the peanut butter and continue to mix until well combined. Blend in the egg, vanilla, baking soda, and salt, scraping down the side of the bowl as needed.

In a separate bowl, whisk together the brown rice flour, sorghum flour, oat flour, tapioca starch, and sweet brown rice flour. Combine with the peanut butter mixture and continue to mix until a dough forms.

Using a cookie scoop, measure out dough into 3- to 4-tablespoon (84 to 112 g) balls and place 2 inches (5 cm) apart on the prepared baking sheet(s). Dip a fork in the extra sweet brown rice flour and press down on each cookie twice to create a crisscross pattern on the top.

Bake for 12 to 14 minutes, until the cookies are golden and the edges are slightly firm. Let cool before serving. Store in an airtight container at room temperature for 3 to 4 days, or the freezer for longer storage.

STRAWBERRY BOURBON CLAFOUTI

I wanted to find ways to use brown rice flour by itself,
because I always have brown rice flour around my house but do not always have
the other gluten-free flours. This clafouti is a delicious, custard-like dessert or breakfast.
A traditional clafouti calls for cherries, but I like to mix the fruit up
by trying strawberries, peaches, or even apples.

YIELD: 6 TO 8 SERVINGS

1 cup (235 ml) 2% or whole milk
3 large eggs
½ cup (72 g) brown rice flour
⅓ cup (80 ml) maple syrup
1 tablespoon (15 ml) bourbon
1 teaspoon vanilla extract
¼ teaspoon sea salt
2 cups (240 g) sliced strawberries

Preheat the oven to 350°F (180°C, or gas mark 4). Grease a 2½-quart (2.4-L) casserole dish with butter.

In a blender, combine the milk, eggs, flour, maple syrup, bourbon, vanilla, and salt. Blend until smooth.

Layer the sliced strawberries in the bottom of the dish. Carefully pour the batter over the strawberries.

Bake for 40 to 45 minutes, until the custard is set and is beginning to brown. Serve warm. Although clafouti is best eaten warm, it can be kept, covered, in the refrigerator and reheated in a 350°F (180°C, or gas mark 4) oven for 10 to 12 minutes.

AMARANTH

The first time I cooked with amaranth, I was confused and not sure I would give the grain a second chance. I assumed it would act like any other grain and cook up with a rice-like consistency. Instead, I ended up with porridge, which was okay but not what I wanted. I shoved amaranth to the back of my cupboard and forgot about it until I stumbled on a delicious-sounding porridge using amaranth. I pulled the grain from the depths of my cupboard, already knowing how wonderful a porridge it could make, and started making this hearty treat for breakfast.

Amaranth falls into the pseudo-grain category, because it is actually the seed of a beautiful flowering plant, which is dried to obtain its seeds. Amaranth is a native crop of Peru and dates as far back as the Incas; it is eaten extensively in Central and South America. The small seed is packed full of nutrients, including calcium and iron, and it contains high levels of protein, making it a perfect addition to any meal.

Beyond the delicious breakfast porridge, amaranth can also be popped, similar to popcorn, and used as a light and crunchy cereal or addition to granola. I also enjoy adding amaranth, along with quinoa, oats, and millet, to the Basic Honey Wheat Bread on page 60 to create a delightful multigrain bread.

The grain is tiny, usually a pale cream color with a few seeds that are more of a brownish black. Amaranth is unique in that as both a grain and a flour it has an earthy, almost molasses or malty type of flavor. Amaranth is available in most health food stores, prepackaged or in the bulk bin.

Amaranth Flour

Amaranth's good nutritional profile has me sneaking amaranth flour into quite a bit of my gluten-free and gluten baking. The flour is a tan color that comes with a unique taste that pairs well with sorghum and oat flours, both of which can help offset the slightly bitter aftertaste amaranth flour can lend to baked goods. Maple syrup and molasses also make good flavor accompaniments.

If the taste of amaranth flour doesn't bother you, then the flour can be used for 100 percent of other flours in recipes that do not require gluten, such as muffins and pancakes. The end texture can be slightly denser, but the overall dish is still delicious.

Grinding amaranth works best in a high-speed blender or coffee grinder. The tiny seeds can be ground in a grain mill, but they can sometimes slip through unground, creating a chunky flour, or they can clog the grinder and stop the grain mill. If using a grain mill, pour the amaranth in a slow, steady stream and sift the ground flour.

Weights and Measurements

1 cup amaranth = 180 g

1 cup amaranth flour = 120 g

1 cup (180 g) amaranth = about 1½ cups (180 g) amaranth flour

BLUEBERRY AMARANTH MUFFINS

If my day has gotten off to a slightly rocky start, a midmorning muffin and cup of coffee can sometimes help turn it around. These muffins are gluten-free but come out of the oven moist and delicious. If you don't have fresh berries, toss frozen blueberries with 1 tablespoon (7.5 g) amaranth flour before adding to the batter and then work quickly or else the batter will turn blue.

YIELD: 12 MUFFINS

1½ cups (180 g) amaranth flour
¾ cup (74 g) oat flour
½ cup (68 g) sorghum flour
½ cup (64 g) arrowroot
1 teaspoon ground cinnamon
1 teaspoon baking powder
½ teaspoon baking soda
½ teaspoon sea salt
½ teaspoon ground ginger
¾ cup (175 ml) 2% or whole milk
¾ cup (175 ml) maple syrup
2 large eggs
½ cup (112 g) butter, melted
1½ cups (210 g) blueberries

Preheat the oven to 350°F (180°C, or gas mark 4). Line a standard 12-cup muffin pan with liners.

In a large bowl, stir together the flours, arrowroot, cinnamon, baking powder, baking soda, salt, and ginger. In a separate bowl, whisk together the milk, maple syrup, eggs, and butter. Pour the wet ingredients into the dry ingredients and stir until just combined.

Fold in the blueberries, being careful not to stir the batter too much. Divide batter among the 12 muffin cups. Bake for 20 to 24 minutes, until the muffins spring back when pressed. Let cool slightly before removing from the muffin pan. Store the muffins in an airtight container at room temperature for 1 to 2 days, or freeze for longer storage.

CINNAMON RAISIN
AMARANTH PANCAKES

If you were to look into the archives of my blog, you would notice
a repeating trend of food that I post about: tacos, spring rolls, and pancakes.
These amaranth pancakes are the perfect way to mix up the pancake routine. If you
do not like raisins, try adding a few chocolate chips instead, or simply
leave the raisins out and top with your favorite fruit.

YIELD: 6 PANCAKES

½ cup (60 g) amaranth flour

¼ cup (24 g) oat flour

¼ cup (30 g) buckwheat flour

2 teaspoons ground cinnamon

1 teaspoon baking powder

¼ teaspoon sea salt

½ cup (64 g) raisins

2 large eggs

½ cup plus 2 tablespoons (150 ml) 2% or
whole milk, plus extra, if needed

1 tablespoon (15 ml) walnut oil or melted
butter

1 tablespoon (15 ml) maple syrup

Preheat the oven to 200°F (100°C, or very low).

In a medium bowl, combine the flours. Stir in the cinnamon, baking
powder, salt, and raisins. In a separate bowl, whisk together the eggs,
milk, walnut oil, and maple syrup. Pour the wet ingredients into the dry
ingredients and stir until combined.

Heat a lightly greased griddle or large skillet over medium heat. Scoop
a scant ¼ cup batter onto the hot skillet, repeating 2 or 3 times without
crowding the pancakes. If the batter thickens to the point where it
doesn't pour, add 1 to 2 tablespoons (15 to 30 ml) more milk to thin it.
Cook the pancakes on each side for 2 to 3 minutes, until golden. Repeat
with the remaining batter. Stack the pancakes on a baking sheet and
keep in the oven until ready to serve.

🖉 These pancakes are thick and fluffy. If you are more of a thin pancake person, use
3 to 4 tablespoons (45 to 60 ml) more milk to thin out the batter.

AMARANTH-CRUSTED ASPARAGUS FRIES

Asparagus is one of my favorite treats of spring, and for
the most part, I simply roast it with a bit of olive oil and salt. However, I sometimes
like to change up my routine, and these asparagus fries are a fun spin
on a fried vegetable appetizer.

YIELD: 3 OR 4 SERVINGS

¾ cup (90 g) amaranth flour

2 teaspoons garlic powder

2 teaspoons dried oregano

¾ teaspoon sea salt

½ teaspoon freshly ground black pepper

½ teaspoon smoked paprika

2 large eggs

2 tablespoons (30 ml) heavy cream

3 to 4 tablespoons (42 to 56 g) coconut oil, divided

½ pound (225 g) asparagus spears, bottoms trimmed

In a shallow dish, combine the amaranth flour, garlic powder, oregano, salt, pepper, and smoked paprika. Stir until combined. In a separate shallow dish, whisk together the eggs and heavy cream.

In a large frying pan with tall sides over medium-low heat, heat 2 tablespoons (28 g) of the coconut oil. Dip 1 asparagus spear in the egg mixture and then in the flour mixture. Repeat once more, to create a double coating, and then place in the pan. Repeat with the remaining asparagus. Fry for 3 to 4 minutes, rotating occasionally with tongs, until the asparagus is golden. Add more coconut oil to the pan as needed.

Occasionally I fry up half the batch, then clean out the skillet and start fresh so that the outside shell stays golden and fresh. Excess amaranth often falls off the asparagus, leaving the frying oil a bit dirty.

MILLET

Millet is quite possibly one of my favorite gluten-free grains. The slightly nutty yet unimposing flavor is a perfect complement to many dishes, and millet works well as a whole grain or flour. Millet has a rich history, dating back thousands of years as a staple in the dry, arid climates of Africa and Asia. In the United States, millet's popularity has primarily been for its use in birdseed rather than for human consumption. In recent years, however, thanks to a surge in overall grain popularity, millet is finding its way into more recipes and meals.

Though it is cooked and consumed as a grain, the label "grain" can be mislead-ing, because millet is actually a seed from the grass family. Millet comes in many shapes and colors, with the primary varieties being pearl, proso, foxtail, and finger. In the United States, proso millet is grown for its use in birdseed and for human consumption, while in most other countries, pearl millet is the most popular. The flavor and texture don't differ greatly from each variety, because all varieties have a slightly nutty flavor.

Millet is sold in many specialty stores, both prepackaged and in bulk bins. The seeds are the size of a small bead and yellow in color. Look for hulled millet instead of pearled, as it has only the outermost shell removed, maintaining most of the fiber. Along with fiber, millet is a good source of magnesium, manganese, and phosphorus.

Millet Flour

Millet grinds into a fine and delicate flour that creates a crumbly texture in baked goods. The flavor of the flour is much like the grain itself: subtle and slightly nutty, with a bit of sweetness. For extra flavor, roast the millet in a dry skillet over medium heat for 8 to 10 minutes. Let it cool and grind as normal.

An electric or hand mill works best to grind millet into flour. A coffee grinder, blender, or food processor will work as well, but sifting the flour is a must to remove any unground pieces.

Millet flour can work in non-yeasted recipes but often requires a little help from either eggs or a starch. To use in yeasted recipes, perhaps as a substitute for wheat flour, replace only one-third of the wheat flour with millet flour, since millet does not contain the gluten that is needed to help hold the shape of tradi-tional baked goods.

Weights and Measurements

1 cup hulled millet = 180 g

1 cup millet flour = 120 g

1 cup (180 g) hulled millet = 1½ cups (180 g) millet flour

MILLET "POLENTA" WITH CHICKPEAS AND SPICY TOMATO CHUTNEY

I love a good, comforting, creamy polenta, and millet makes an ideal substitute for cornmeal while adding a delightful nutty flavor.

YIELD: 2 MAIN-DISH OR 4 SIDE-DISH SERVINGS

For the chutney:

1 tablespoon (15 ml) olive oil

1 shallot, minced

1 clove garlic, minced

1½ teaspoons minced fresh ginger

2 pounds (910 g) Roma tomatoes, quartered

1 teaspoon brown mustard seeds

¼ teaspoon crushed red pepper

2 tablespoons (30 ml) fresh lemon juice

2 tablespoons (40 g) honey

⅓ cup (50 g) golden raisins

For the "polenta":

1 tablespoon (15 ml) olive oil

1 cup (150 g) finely diced onion

2 cups (475 ml) water

2 cups (475 ml) low-sodium vegetable broth

1¼ cups (150 g) millet flour

½ teaspoon sea salt

1 cup (240 g) cooked chickpeas, rinsed and drained if using canned

To make the chutney: Heat a medium saucepan over medium heat and add the olive oil. Add the shallot and cook until translucent, 4 to 5 minutes. Stir in the garlic and ginger, and cook for 1 to 2 more minutes. Add the tomatoes along with the mustard seeds, crushed red pepper, lemon juice, honey, and raisins. Cook until the tomatoes reduce slightly and the flavors come together, 30 to 45 minutes.

To make the "polenta": Heat the olive oil in a large pot. Add the onion and cook until translucent, 4 to 5 minutes. Stir in the water and vegetable broth and bring to a boil. Whisk in the millet flour and salt. Continue to cook, whisking frequently, for 10 to 12 minutes, until the polenta thickens. Remove from the heat and let sit for 1 minute. Whisk one more time.

Heat the chickpeas in a skillet or saucepan over medium-low heat.

To serve, place the millet polenta on plates and top with the tomato chutney and chickpeas.

For a more textured polenta, grind millet in a spice or coffee grinder, leaving some millet in a cracked state. Increase cooking time to 15 minutes and add more vegetable broth as needed to keep a consistency that can be whisked.

MILLET WAFFLES WITH MAPLE STRAWBERRY SAUCE

Waffles with millet flour are the perfect texture for me:
crisp on the outside with a light and fluffy inside. Depending on the season, substitute
blueberries or blackberries for the strawberries. If you want to keep the waffles crisp
before serving, preheat your oven to 200°F (100°C, or very low) and, as the waffles are
done, lay them on a baking sheet and place in the oven until ready to serve.

YIELD: 2 BELGIAN OR 4 REGULAR WAFFLES

For the sauce:

3 cups (360 g) quartered strawberries

3 tablespoons (45 ml) maple syrup

½ teaspoon ground cinnamon

For the waffles:

¾ cup (90 g) millet flour

¼ cup (25 g) oat flour

¼ cup (30 g) buckwheat flour

1 teaspoon baking powder

¼ teaspoon fine sea salt

2 large eggs

¼ cup (60 g) whole-milk plain
   Greek yogurt

3 tablespoons (45 ml) 2% milk

2 tablespoons (30 ml) maple syrup

2 tablespoons (30 ml) walnut oil

½ teaspoon vanilla extract

To make the sauce: Combine the strawberries, maple syrup, and cinnamon in a small saucepan over medium-low heat. Let simmer while you make the waffles.

To make the waffles: Preheat a waffle iron according to the manufacturer's instructions.

In a medium bowl, stir together the flours, baking powder, and salt. In a separate bowl, whisk together the eggs, yogurt, milk, maple syrup, walnut oil, and vanilla. Pour the wet ingredients into the dry ingredients and stir until combined.

For full, Belgian-style waffles, pour half the batter into the waffle iron and follow your waffle maker's instructions until each side is golden brown and crispy. For smaller waffles, pour in one-quarter of the batter and follow the waffle maker's instructions. Repeat with the remaining batter. Serve each waffle with the strawberry topping.

MILLET POTPIES

Occasionally my grandmother would heat a frozen
potpie for a special lunchtime treat. Although I don't eat much meat anymore,
I often feel a bit nostalgic for this savory pie. The millet flour creates a crumbly texture
that pairs perfectly with the creamy butternut squash filling. Don't be worried if the
biscuits seem a bit sticky, as the dough will spread perfectly over the filling!

YIELD: 4 SERVINGS

For the filling:

1 tablespoon (15 ml) olive oil

1 medium (200 g) onion, diced

½ cup (120 ml) water

3 tablespoons (5 g) chopped fresh
rosemary

1 teaspoon freshly ground black pepper

½ teaspoon sea salt

6 cups (750 g) cubed butternut squash
(from about a 3-pound or 1.4-kg squash)

¼ cup (30 g) millet flour

2⅔ cups (640 ml) low-sodium vegetable
broth

For the crust:

1 cup (120 g) millet flour

2 tablespoons (24 g) arrowroot

2 teaspoons baking powder

½ teaspoon sea salt

⅛ teaspoon baking soda

¾ cup (24 g) grated Parmesan, divided

¼ cup (56 g) cold butter, cubed

¼ cup (60 ml) whole milk

1 tablespoon (20 g) honey

1 large egg

1 tablespoon (15 ml) water

Preheat the oven to 375°F (190°C, or gas mark 5).

To make the filling: Heat the olive oil in a medium pot over medium heat. Add the onion and cook for 4 to 5 minutes, until translucent. Add the water, rosemary, pepper, salt, and squash. Cook, stirring, for 5 minutes.

Sprinkle the millet flour over the mixture and cook for 1 to 2 minutes, letting the flour taste cook out. Pour in the vegetable broth and stir well. Continue to cook until the mixture has thickened. Remove from the heat and divide among four 10-ounce ramekins.

To make the crust: In a food processor or in a bowl, combine the millet flour, arrowroot, baking powder, salt, baking soda, and ½ cup (16 g) of the Parmesan. Pulse or stir until combined. Next, cut in the butter, either using the food processor or by hand with a pastry blender or 2 knives, until the dough is in pea-size pieces. Pour in the milk and honey and continue to pulse or stir until a dough begins to form.

Divide the dough into 4 balls. With floured hands, pat each ball into a ½-inch-thick (1.2 cm) circle and lay over the filling. In a small bowl, whisk together the egg and water. Brush the tops of the potpies with the egg wash and sprinkle remaining the ¼ cup (8 g) Parmesan on top. Set the ramekins on a baking sheet.

Bake for 20 to 22 minutes, until the crust is golden and the potpies are bubbly.

QUINOA

I discovered quinoa only recently, but the little seed has quickly become a favorite in my house. Quinoa is grown in the Andes region of South America, primarily in Peru and Bolivia. It has a rich history dating back as one of the two main staple foods for the Incas and has continued to remain a staple for South American cultures. In the past few years, quinoa has become quite popular outside of South America. The small grain is praised for its nutty flavor, quick cooking time, and good nutrition—it contains all the essential amino acids (making it a complete protein), potassium, and vitamin E.

Although quinoa is a seed, it is classified as a grain because of its uses in cooking and its grain-like qualities. While there are more than 120 varieties of quinoa, the main varieties sold prepackaged and in bulk bins are white, red, and black. There are subtle differences among them, such as more calories and protein in the red, while the white tends to be a bit creamier when cooked. I keep all three in my kitchen for different uses.

Quinoa can often have a bitter taste, which is caused by saponin, a naturally occurring chemical that coats the quinoa during the growing phase. Even though prepackaged quinoa is rinsed before being sold, it is useful to rinse the quinoa again before use to help mitigate this bitter taste.

Quinoa Flour

Quinoa flour has a hearty texture that makes it a wonderful addition to baked goods and breads. Substitute quinoa flour for half the all-purpose flour in recipes. Quinoa also makes a wonderful binding agent, like in the Sweet Potato and Quinoa Flour Patties with Poached Eggs (page 104), or a crispy crust, such as in the Quinoa Flour–Crusted Cauliflower Steaks with Roasted Tomato Sauce (page 103).

I grind whichever color quinoa I have on hand because the colors have little effect on the outcome of the flour. Red quinoa flour is a bit darker than wheat flour, while white quinoa flour is lighter in color.

To make quinoa flour using a blender or food processor, first rinse the quinoa under cold water. Working in small batches, add the quinoa to a hot skillet and heat until it starts jumping around. Remove from the heat, and swirl around until the quinoa pops, just like popcorn. Let cool and then grind. Sift the flour to remove any small bits of seed still remaining.

To grind quinoa in a grain mill or coffee grinder, rinse the quinoa under cold water and then spread onto a jelly-roll pan or other pan with a lip. Dry it in a 350°F (180°C, or gas mark 4) oven for 10 minutes for a neutral flavor and 15 minutes for a slightly nuttier flavor.

Weights and Measurements

1 cup quinoa = 170 g

1 cup quinoa flour = 112 g

1 cup (170 g) quinoa = 1½ cups (168 g) quinoa flour

CHEDDAR JALAPEÑO QUESADILLAS WITH QUINOA TORTILLAS

Quesadillas tend to be my quick lunch or dinner meal,
and these quinoa tortillas add just a hint of earthiness that pairs perfectly
with the jalapeños. I tend to use whatever cheese I have on hand
at the moment, but I've found that Cheddar works the best.

YIELD: TWO 8-INCH (20-CM) QUESADILLAS

For the filling:

1 tablespoon (15 ml) olive oil

3 or 4 medium jalapeño chile peppers, sliced (keep seeds for heat)

1 medium (160 g) yellow onion, sliced

For the tortillas:

½ cup (56 g) quinoa flour

¼ cup plus 2 tablespoons (45 g) millet flour

¼ cup (36 g) brown rice flour

2 tablespoons (16 g) tapioca starch

2 tablespoons (16 g) cornstarch

½ teaspoon sea salt

½ teaspoon baking powder

½ cup (118 ml) water

2 tablespoons (30 ml) olive oil

6 ounces (170 g) Cheddar, shredded

To make the filling: Heat the olive oil in small skillet over medium-low heat. Add the jalapeños and onion. Cook for 8 to 10 minutes, stirring frequently, until tender.

To make the tortillas: In a large bowl, whisk together the flours, starches, salt, and baking powder. Pour in the water and olive oil, mixing until the dough comes together into a ball. Let the dough sit for 5 minutes.

Preheat a griddle or large skillet over medium heat. Generously flour your work surface with quinoa flour. Divide the dough into 4 balls. Roll each out into an 8-inch (20-cm) circle. Transfer the tortillas to the griddle and cook on each side until lightly golden and forming air pockets, 1 to 2 minutes. Do not let the tortillas cook for too long, as they crisp quickly. Transfer to a plate.

To assemble the quesadillas, return a tortilla to the griddle. For a full quesadilla, spread half of the jalapeño mixture and cheese over a tortilla. Cover with a second tortilla. For half quesadillas, spread one-quarter of the jalapeño mixture and cheese over one-half of each tortilla and fold the other half over the top.

Cook the quesadillas on each side for 3 to 5 minutes, or until the tortillas are crispy and the cheese has melted.

These tortillas are fragile. Have patience when rolling them out and transferring them to a pan. I highly recommend using a bench scraper to aid in moving the tortillas. Also, err on the side of too much flour when rolling them out.

QUINOA FLOUR—CRUSTED CAULIFLOWER STEAKS WITH ROASTED TOMATO SAUCE

As a vegetarian, I have had a few occasions when I miss having
a "steak and potatoes" meal. Although this isn't the kind of steak that first comes
to mind, these cauliflower steaks are one of my go-to hearty vegetarian meals.
The addition of the quinoa flour creates a nice crispy crust that adds
a wonderful hint of nuttiness to the overall dish.

YIELD: 2 SERVINGS

For the sauce:

4 cups (560 g) cherry or grape tomatoes,
halved

1 medium (160 g) white onion, chopped

2 cloves garlic, minced

2 tablespoons (30 ml) olive oil

¼ cup (10 g) fresh basil leaves, julienned

For the steaks:

1 large (800 g) head cauliflower

2 large eggs

¼ cup (60 ml) heavy cream

½ cup (56 g) quinoa flour

½ teaspoon sea salt

½ teaspoon freshly ground black pepper

2 tablespoons (30 ml) olive oil, divided

1 to 2 ounces (28 to 55 g) goat cheese,
crumbled

2 to 3 tablespoons (5 to 8 g) fresh basil
leaves, julienned

Preheat the oven to 400°F (200°C, or gas mark 6). Lightly grease a
baking sheet or line it with parchment paper.

To make the sauce: In a large bowl, toss the tomatoes, onion, and garlic
with the olive oil. Spread in a single layer in a roasting pan or on a
baking sheet with sides. Roast until the tomatoes are starting to brown
and the onions are starting to caramelize, 30 to 35 minutes. Transfer to
a bowl, add the basil, and pulse until a sauce forms. Set aside.

To make the steaks: Strip away any excess greens on the head of cauliflower. With the stem side down on the cutting board, cut two ½-inch
(1.2-cm) thick steaks from the center of the cauliflower, reserving florets
as they fall off.

In a shallow dish, whisk together the eggs and cream. In a separate shallow dish, combine the quinoa flour, salt, and pepper. Coat the cauliflower
steaks in the egg mixture, then carefully transfer to the quinoa flour and
coat. Repeat the process, coating with the egg and quinoa flour one
more time to create a double crust.

Heat half the olive oil over medium-low heat in a skillet big enough to
hold 1 cauliflower steak. Fry the steak for 3 to 4 minutes on each side,
until browned and crispy. Transfer the steak to a baking sheet. Heat the
other half of the olive oil and repeat with the second steak. Bake the
steaks for 15 minutes, until tender.

Serve the steaks with the roasted tomato sauce, topped with the goat
cheese and a sprinkle of basil.

SWEET POTATO AND QUINOA FLOUR PATTIES WITH POACHED EGGS

I eat a lot of eggs, and I am constantly looking for new ways
to go beyond my breakfast scramble. This dish is hearty enough to make
a satisfying lunch or dinner. The quinoa flour adds extra nutrition to
the patties, while the egg helps round out the meal.

YIELD: 2 SERVINGS

1 (1-pound, or 455 g) sweet potato

1 tablespoon (6 g) cumin seeds

1 tablespoon (5 g) coriander seeds

¾ cup (84 g) quinoa flour

½ teaspoon sea salt

1 clove garlic, minced

3 large eggs, divided

¼ cup (4 g) fresh cilantro, minced

1 tablespoon (15 ml) olive oil

Fresh cilantro leaves, for garnish

Preheat the oven to 400°F (200°C, or gas mark 6).

Pierce the sweet potato with a fork and bake until tender, 40 to
50 minutes. Let cool slightly.

Toast the cumin and coriander seeds for 2 to 3 minutes in a small dry
skillet over medium heat. Remove and grind with a mortar and pestle
or in a spice grinder.

Scoop out the baked sweet potato flesh into a large bowl. Add the
ground spices, quinoa flour, salt, garlic, 1 egg, and the cilantro.
Stir until the mixture comes together.

Heat the olive oil in a large skillet over medium heat. Wet your hands
and form the sweet potato mixture into four ½-inch (1.2-cm)-thick patties.
Place in the skillet and cook on each side until golden, 4 to 5 minutes
per side.

While the patties are cooking, bring a large pot of water to a simmer.
Lower the heat slightly so that the water is still. Crack each remaining
egg into separate small bowls and then carefully slip the eggs into the
hot water. Cook until the egg whites have set and the yolk is desired consistency, 4 minutes for a softer yolk, 6 minutes for a slightly firmer yolk.
Remove the eggs with a slotted spoon and drain slightly on paper towels.

To serve, place 2 patties on each plate, and top with a poached egg and
a few cilantro leaves.

I love using whole spices for their fresh flavor. However, ground cumin and
coriander can be used in this recipe if needed. Replace each seed spice with
2 teaspoons of each ground spice, and skip the toasting step.

OATS

I keep two large containers on my counter: one full of brown rice and the other full of rolled oats. Brown rice is my dinner staple, and oats are my breakfast standby. Even before I started grinding my own wheat flour, I was grinding oats in my food processor for pancakes and waffles or to add to my whole wheat bread. I have always loved the sweet taste and moistness that oats give to baked goods.

Navigating the whole-grain spectrum can be confusing when some variations of a particular grain aren't considered "whole" due to a processing procedure that strips away nutrients. Oats, however, are a little easier to understand in that most versions (including rolled oats, whole oat groats, and steel-cut oats) are considered whole grains with the nutrients still intact. All oat products are created from oat groats, which are oat kernels with the hull removed. Steel-cut oats and Scottish oats are oats that have been cut or ground into smaller, quicker-cooking pieces from the oat groats, while rolled oats, such as old-fashioned oats, come from steaming and flattening the oat groat.

I have both rolled oats and oat groats on hand for breakfast and grain salads. A variety of oats can be found at a health food store, prepackaged and in bulk bins, and at supermarkets, usually in the form of rolled oats or steel-cut oats. If you're eating gluten-free for health reasons, use caution when buying oats, because many of the different kinds of oats can become contaminated with gluten during processing. (It pays to research brands before buying.)

Oat Flour

Besides my love of oatmeal for breakfast, I also enjoy oat flour and use it often, particularly by itself. It is easy to create from any of the available oat products and adds a lovely sweetness and nuttiness to the final dish. Oat flour also makes a great addition to a mix with other flours because it adds extra moisture in addition to sweetness.

To make oat flour from rolled oats, simply process them in a blender or food processor until they reach flour consistency. If you're looking for a smooth flour consistency, run the oats in the food processor, sift through a sieve, and return any pieces of oat to the food processor, continuing until nearly all the oat pieces are ground into flour. For oat groats or steel-cut oats, use a grain mill, high-powered blender, or coffee mill.

Because of the higher fat content in oats, the flour can go rancid fairly quickly after being ground. I grind only what I need, and if I have any left over, I keep it in the refrigerator for up to 2 weeks.

Weights and Measurements

1 cup oat groats = 180 g

1 cup rolled oats = 100 g

1 cup oat flour = 100 g

1 cup (180 g) oat groats = about 1¾ cups (175 g) oat flour

1 cup (100 g) rolled oats = 1 cup (100 g) oat flour

CRANBERRY OAT COOKIES

These cookies came about one day when I was on a mission
to make chocolate chip oat cookies. I searched my cupboards only to find that
I didn't have any chocolate chips, but I did have dried cranberries and pecans.
The combination makes for a unique and wonderfully sweet cookie. By the time I finished
my first cookie, I had forgotten that I had originally wanted chocolate chips!

YIELD: 18 COOKIES

½ cup (112 g) butter, softened

½ cup (115 g) packed brown sugar

½ cup (100 g) organic cane sugar

1 teaspoon baking soda

½ teaspoon sea salt

1½ teaspoons vanilla extract

2 large eggs

3 cups (300 g) oat flour

1 cup (120 g) dried cranberries

½ cup (50 g) toasted pecan pieces

Preheat the oven to 375°F (190°C, or gas mark 5). Line a baking sheet (or 2) with parchment paper.

In a large bowl with a hand mixer or a stand mixer fitted with the paddle attachment, cream the butter, brown sugar, and cane sugar. (Alternatively, mash together the butter and sugars with a wooden spoon until completely combined.) Continue to mix until well combined and there are no chunks of butter. Add the baking soda, salt, vanilla, and eggs, scraping down the side of the bowl as needed.

Add the oat flour and continue to mix until a dough forms. Once the ingredients are well combined, stir in the dried cranberries and pecan pieces until evenly distributed. Scoop out the dough in roughly 2-ounce (55-g) balls and set on the prepared baking sheet. Press down slightly with the back of a spoon covered with oat flour. Bake for 12 to 14 minutes, until the cookies are golden on top. Let cool on the baking sheet until the cookies have set. Store the cookies in a airtight container at room temperature for 3 to 4 days, or freeze for extended storage.

✎ This dough is wet. I find it's easier to use an ice cream scoop for the dough and then use the spoon covered in flour to press down the cookies.

BERRY COBBLER WITH OAT DUMPLINGS

During the summer, I am fairly certain that I eat my weight
in berries: I can't seem to get enough. This cobbler is the perfect way to highlight
all these summer fruits in a delicious, light dessert. I've drastically cut down on the
sugar here, compared with other cobbler recipes, but if you use ripe berries, you won't
miss it, because the berries are extremely sweet on their own. If you like, serve with
a scoop of vanilla ice cream or a drizzle of heavy cream.

YIELD: 4 TO 6 SERVINGS

For the filling:

6 cups (750 g) mixed berries (strawberries, blackberries, raspberries, blueberries)

⅓ cup (80 ml) maple syrup

¼ cup (50 g) organic cane sugar

1 teaspoon ground cinnamon

3 tablespoons (34 g) cornstarch

2 tablespoons (30 ml) water

1 tablespoon (15 ml) fresh lemon juice

For the topping:

2 cups (200 g) oat flour

1 tablespoon (4.6 g) baking powder

¾ teaspoon sea salt

¼ cup plus 2 tablespoons (84 g) butter, cut into pieces

1 large egg

2 tablespoons (30 ml) whole milk

2 tablespoons (30 ml) maple syrup

Preheat the oven to 375°F (190°C, or gas mark 5).

To make the filling: Combine the berries, maple syrup, sugar, and cinnamon in a pot and bring to a boil, cooking for 3 to 4 minutes as the berries expel their juice. In a small bowl, whisk together the cornstarch, water, and lemon juice until smooth. Pour into the boiling berry mixture and cook for 2 to 3 more minutes, stirring frequently, until the mixture starts to thicken. Pour into a 9-inch (23-cm) round baking dish.

To make the topping: In a large bowl, combine the oat flour, baking powder, and salt. Cut in the butter, using a pastry blender, 2 knives, or your fingers, until the dough is in pea-size pieces. In a small bowl, whisk together the egg, milk, and maple syrup. Pour the wet ingredients into the dry ingredients and stir until a dough forms.

Break off golf ball–size pieces of the dough, lightly pat into a circle, and place over the berry mixture. Continue with the remaining dough. Bake for 25 to 30 minutes, until the filling is bubbling and the topping is golden.

OATCAKES WITH
MAPLE BANANAS

I first discovered the concept of using oat flour in pancakes when
I was craving pancakes one morning but I didn't have any flour. I ground rolled
oats and went to work. The oatcakes turned out so light and perfect that they
made their way into my permanent breakfast rotation.

YIELD: 8 TO 10 PANCAKES

For the oatcakes:

1 cup (100 g) oat flour

1 teaspoon baking powder

½ teaspoon sea salt

2 large eggs

½ cup (112 g) cup whole-milk plain
   Greek yogurt

2 tablespoons (30 ml) walnut oil or melted
   butter

1 tablespoon (15 ml) maple syrup

For the bananas:

2 large bananas, cut into ½-inch (1.2-cm)
   slices

4 teaspoons maple syrup

½ teaspoon ground cinnamon

To make the oatcakes: In a medium bowl, stir together the oat flour, baking powder, and salt. In a separate bowl, whisk together the eggs, yogurt, oil or butter, and maple syrup. Pour the wet ingredients into the dry ingredients and stir until just combined.

Heat a griddle or skillet over medium heat and lightly grease with butter or oil if need be. Pour ¼ cup (48 g) of the batter onto the heated surface and slightly push around with a spoon or the back of the measuring cup to form a pancake. Repeat 2 or 3 more times, without crowding the skillet. Let cook until bubbles begin to appear and pancake is set, 2 to 3 minutes. Flip and cook for 2 to 3 minutes more. Set aside cooked pancakes on an ovenproof plate and keep in a 200°F (100°C, or very low) oven until ready to serve. Continue with the remaining batter.

To make the bananas: In a skillet over medium heat, combine the bananas with the maple syrup and cinnamon. Cook, stirring occasionally, until the bananas are partially caramelized. Serve over the pancakes.

This pancake batter is a bit thicker than normal, which is the reason for pushing the pancake into a round shape on the cooking surface. Also, do not get impatient and try to flip the pancake earlier than directed, because it won't hold together.

CORN
(POPCORN)

When I first received my grain mill, I felt like a kid again, thinking up fun experiments about what I could grind with the machine. I scoured the Internet for facts about what would and wouldn't work with the grinder, and that's when I stumbled on corn. I was enthralled with the fact that I could grind popcorn kernels into homemade cornmeal.

The type of corn used for popcorn is characterized by having a hard outside that seals in a very starch-filled inside. When popcorn is heated, the moisture that has been sealed inside the corn kernel releases and pops. Popcorn has a rich history, dating back thousands of years to Peru and spreading through South and North America, becoming an extremely popular snack.

Popcorn can be found in grocery stores and health food stores. Corn has received somewhat of a bad reputation in recent years, what with factory farming, pesticides, and genetic modification. As a result, I tend to buy organic popcorn, local when possible.

Corn Flour

The versatility of flour made from popcorn kernels is rather amazing. I always keep cornmeal on hand to whip up a batch of corn bread or polenta, to use it for pizza, or to add it to breads for extra texture. The flavor of home-ground cornmeal is milder than that of store-bought, making it a perfect canvas for adding herbs and spices.

Proceed with caution when grinding popcorn kernels, because of the hard outer shell. If you own a grain mill, check the manual to determine whether the machine can handle the popcorn. A high-powered blender will work, but sift the cornmeal once it's done, just in case a shell didn't grind all the way. For smaller batches, a sturdy coffee grinder will work as well. Other types of corn can be ground into flour, such as dent corn, which is field corn grown normally for use in processed foods or for livestock, or other varieties of flint corn, such as flour corn.

Weights and Measurements

1 cup popcorn = 204 g

1 cup corn flour = 136 g

1 cup (204 g) popcorn = 1½ cups (204 g) corn flour

Sweet Corn Muffins

During the winter months, I enjoy corn bread smothered with
hot vegetarian chili. Although these muffins don't have the exact corn
bread taste, they are perfect in a pinch and make a wonderful addition to
any meal. Occasionally I add 1 cup (115 g) Cheddar cheese and a minced
jalapeño to the batter for extra flavor.

YIELD: 12 MUFFINS

1 tablespoon (15 ml) olive oil

½ medium (80 g) red onion, minced

Kernels from 1 medium ear corn, or
1¾ cups (300 g) frozen corn kernels,
thawed

1 cup (136 g) corn flour

½ cup (50 g) oat flour

¼ cup (32 g) arrowroot

2 teaspoons baking powder

½ teaspoon sea salt

2 large eggs

¾ cup (175 ml) whole milk

¼ cup (85 g) honey

¼ cup (56 g) butter, melted and slightly
cooled

Preheat the oven to 350°F (180°C, or gas mark 4). Line a standard
12-cup muffin pan with paper liners or lightly grease with melted butter.

In a skillet over medium heat, heat the olive oil. Add the onion and cook
until fragrant and translucent, 6 to 8 minutes. Add the corn and cook
until the corn is tender, 6 to 7 minutes. Remove from the heat and let
cool slightly.

In a medium bowl, combine the corn flour, oat flour, arrowroot, baking
powder, and salt. In a separate bowl, whisk together the eggs, milk,
honey, and melted butter. Pour the wet ingredients into the dry ingredients and stir until just combined. Fold in the corn mixture.

Divide the batter among the 12 muffin cups, filling each well two-thirds
full. Bake for 12 to 15 minutes, until golden brown and the muffins
spring back when pressed. Let cool slightly, then transfer the muffins to
a wire rack to finish cooling. Store the muffins in an airtight container
at room temperature for 2 to 3 days, or freeze for extended storage.

GRILLED POLENTA WITH ROASTED ZUCCHINI SALSA

I was beyond excited the first time I realized that leftover
polenta held together well enough to grill. Grilling, from a vegetarian perspective,
can sometimes be a bit limited on options. This polenta base is useful beyond
the zucchini salsa. I've had it with grilled green tomatoes, topped with
goat cheese, and tossed with wilted greens. The possibilities are endless.

YIELD: 2 TO 4 SERVINGS

For the polenta:

½ cup (68 g) corn flour

2 cups (470 ml) water

½ teaspoon garlic powder

¼ teaspoon sea salt

For the salsa:

1 medium (120 g) zucchini, cubed

½ medium (60 g) onion, diced

¼ medium (30 g) red pepper, diced

1 tablespoon (9 g) minced serrano chile
pepper (optional)

1 tablespoon (15 ml) olive oil

¼ teaspoon sea salt

2 tablespoons (2 g) chopped fresh cilantro

1 tablespoon (15 ml) fresh lime juice

Olive oil, for brushing

To make the polenta: In a medium saucepan, bring the corn flour, water,
garlic powder, and salt to a boil. Cook, stirring frequently, until mixture
thickens to a creamy consistency, 3 to 4 minutes. Spoon the mixture into
an oiled 8 × 8-inch (20 × 20-cm) pan. Let the polenta cool completely.

To make the salsa: Preheat the oven to 400°F (200°C, or gas mark 6).
Toss the zucchini, onion, red pepper, and chile (if using) with the olive oil
and salt. Place in a single layer in a roasting pan and roast for 30 minutes, until the zucchini is soft and beginning to char. Toss the zucchini
mixture with the cilantro and lime juice.

When ready to grill the polenta, slice into 2-inch (5-cm) squares and
brush each side with olive oil. Place on a preheated grill and cook on
each side until browned and crispy, 4 to 6 minutes depending on the
heat of the grill. Serve the polenta hot with the zucchini salsa.

This salsa can also be grilled instead of roasted. Slice the zucchini, pepper,
and onion into ½-inch (1.2-cm) slices and brush with olive oil. Grill until lightly
charred and then toss with the cilantro and lime juice. For thicker polenta slices
to grill, double the polenta recipe but use the same size pan.

BRUSSELS SPROUTS CASSEROLE WITH CORN BREAD TOPPING

I have what some may consider a strange love affair with
Brussels sprouts: I can't ever eat too many. This casserole is a hearty side dish that
combines two of my favorite foods: Brussels sprouts and corn bread. Experiment with
different cheese combinations if blue cheese isn't to your liking.

YIELD: 3 OR 4 SIDE-DISH SERVINGS

For the Brussels sprouts:

¾ pound (340 g) Brussels sprouts, stems
  removed and sprouts halved

1 shallot (60 g), sliced

1 tablespoon (15 ml) olive oil

For the sauce:

1 tablespoon (14 g) butter

1 tablespoon (8 g) cornstarch

½ teaspoon sea salt

½ teaspoon freshly ground black pepper

¾ cup (175 ml) whole milk

3 ounces (85 g) blue cheese, crumbled

For the topping:

½ cup (68 g) corn flour

½ teaspoon baking powder

¼ teaspoon sea salt

1 large egg

2 tablespoons (30 ml) olive oil

2 tablespoons (30 ml) whole milk

1 tablespoon (20 g) honey

To make the Brussels sprouts: Preheat the oven to 350°F (180°C, or gas mark 4).

In a bowl, toss together the Brussels sprouts, shallot, and olive oil. Spread in a single layer in a roasting pan and roast for 30 minutes, until the Brussels sprouts are tender.

To make the sauce: When the Brussels sprouts are nearly done, melt the butter in a small saucepan. Whisk in the cornstarch, salt, and pepper, and cook for 1 minute. Whisk in the milk and cook, continuing to whisk, until the mixture comes to a boil and thickens. Remove from the heat and stir in the blue cheese. Once the cheese has melted, combine the mixture with the roasted Brussels sprouts in a 2-quart (1.9-L) baking dish. Bake for 10 minutes.

To make the topping: While the cheesy Brussels sprouts bake, in a medium bowl, combine the corn flour, baking powder, and salt. In a separate bowl, whisk together the egg, olive oil, milk, and honey. Combine the wet and dry mixtures and pour the mixture over the cheesy Brussels sprouts and return the dish to the oven. Bake for another 16 to 20 minutes, until the topping is golden and springs back when pressed.

BUCKWHEAT

As with oat flour, I was eating buckwheat flour before I really understood the wonders it held for gluten-free baking. My favorite waffles include a hefty scoop of buckwheat flour, which adds an unimposing nutty flavor, and the buckwheat crepes on page 120 are perfect for any sweet or savory use. Buckwheat is the perfect bridge to make a meal that is gluten-free that everyone will love by adding both great flavor and nutritional benefits.

Buckwheat is a bit of an imposter in the grains category, because it is not a grain but rather the seed of a fruit that is related to rhubarb. This seed, similar in size to a wheat berry, is easy to identify by its odd triangular shape. Buckwheat is native to the Balkan regions of Europe but spread into Asia, making it a staple in both Eastern European and Asian diets. Prized for being high in protein and fiber, buckwheat also is full of zinc, copper, and manganese.

Buckwheat is primarily sold in two forms: raw or roasted, the latter often labeled kasha. I usually keep raw buckwheat groats for grain salads and pilafs, but I find that roasted buckwheat can make a flavorful flour. Both raw and roasted buckwheat groats can be found prepackaged and in the bulk bins of health food stores.

Buckwheat Flour

Because of my love of buckwheat in baked goods, the majority of the buckwheat groats in my house are ground into flour. The flavor of roasted versus raw buckwheat can affect the flavor of the end product, and I recommend you experiment with both to determine which you like better. For the majority of my recipes, I grind raw buckwheat groats simply because they tend to be what I have on hand.

Buckwheat grinds well in a grain mill, a high-speed blender, or in a coffee grinder for small batches. Similar to oat flour, buckwheat flour can turn rancid rather quickly after grinding. I recommend grinding only what you need, and if you grind extra, refrigerate the flour for up to 2 weeks.

If gluten is not an issue, try switching one-third of the soft wheat flour for buckwheat flour in the rhubarb piecrust on page 68. Buckwheat flour is the perfect partner for soft wheat flour to keep the lightness of pastry recipes. Or sub buckwheat flour for half of the oat flour in the oatcake recipe on page 111 for a wonderful breakfast treat.

Weights and Measurements

1 cup buckwheat groats, raw or roasted = 180 g

1 cup buckwheat flour = 120 g

1 cup (180 g) buckwheat groats = 1½ cups (120 g) buckwheat flour)

Buckwheat Dutch Baby with Maple Raspberries

One of my favorite weekend breakfasts is a Dutch baby. For those not familiar, a Dutch baby is a type of pancake derived from a German pancake that is baked, preferably in a cast iron skillet, and has a dense texture. It is easy to whip up and results in a fantastic, minimal-effort breakfast. The buckwheat flour adds an extra layer of flavor that is perfect for this hearty pancake. Raspberries work well, but feel free to use other berries, depending on the season.

YIELD: 3 OR 4 SERVINGS

For the Dutch baby:

2 large eggs

¼ cup plus 2 tablespoons (90 ml) 2% milk

½ cup (60 g) buckwheat flour

3 tablespoons (24 g) arrowroot

Zest from ½ lemon

⅛ teaspoon sea salt

2 tablespoons (15 ml) maple syrup

1 teaspoon vanilla extract

1 tablespoon (14 g) butter, plus more for serving

For the raspberries:

1½ cups (180 g) raspberries

2 tablespoons (30 ml) maple syrup

Heavy cream, for serving (optional)

Preheat the oven to 400°F (200°C, or gas mark 6).

To make the Dutch baby: Combine the eggs, milk, flour, arrowroot, lemon zest, salt, maple syrup, and vanilla in a blender and run until the batter is smooth.

On the stove top, melt the butter in an 8-inch (20-cm) or 10-inch (25.5-cm) cast-iron or other ovenproof skillet. Swirl around to cover the pan. Remove from the heat and pour in the batter. Place in the oven and bake until golden brown and puffed, 22 to 24 minutes for an 8-inch (20-cm) skillet or 18 to 22 minutes for a 10-inch (25.5-cm) skillet.

To make the raspberries: Combine the raspberries and maple syrup in a small saucepan over medium-low heat. Cook until the raspberries break down into a sauce, 10 to 12 minutes.

Serve the Dutch baby with a smear of butter, the raspberry sauce, and a drizzle of heavy cream, if desired.

✎ This is one recipe in which the arrowroot is optional. Dutch babies, by nature, are dense. The arrowroot helps to lighten the pancake slightly but really isn't missed if left out.

TALEGGIO GRILLED CHEESE WITH SPINACH AND BUCKWHEAT CREPES

I love a good, gooey grilled cheese. However, I sometimes
want the filling to shine a bit more than the bread around it. Crepes make the
perfect bread alternative in these situations—light, easy to make, and perfectly crispy.
If you have never tried Taleggio cheese, I urge you to—it melts beautifully.

YIELD: 2 SERVINGS

For the crepes:

¼ cup (30 g) buckwheat flour

⅛ teaspoon sea salt

1 large egg

3 tablespoons (45 ml) 2% or whole milk

1 tablespoon (15 ml) walnut oil

For the filling:

1 tablespoon (15 ml) olive oil

2 cups (80 g) packed spinach

¼ teaspoon crushed red pepper (optional)

2 ounces (55 g) Taleggio cheese, shredded

To make the crepes: Combine the buckwheat flour, salt, egg, milk, and walnut oil in a bowl; whisk until smooth. Heat an 8-inch (20-cm) skillet over medium-low heat and lightly grease with oil. Pour a scant ¼ cup (36 g) batter into the pan and tilt so that the batter covers the entire pan. Cook for about 30 seconds, until the crepe looks set and the edges pull away from the side of the pan. Flip and cook for another 15 seconds. As the crepes are cooked, layer on a plate, barely overlapping. When finished, make the filling.

To make the filling: Return the skillet to medium-low heat and heat the olive oil. Add the spinach and crushed red pepper, if using. Stir, turn off the heat, cover, and let sit until the spinach is slightly wilted, 3 to 4 minutes.

To make the grilled cheese, add about ¼ cup (37 g) spinach mixture to one half of a crepe. Sprinkle with one-quarter of the cheese. Fold the empty side over the mixture, and then fold in half again to make a triangle. Place in the skillet over medium-low heat, and cook until the cheese is melted and both sides are golden brown, 2 to 3 minutes for each side. Repeat with the remaining crepes and spinach mixture.

Depending on the season, if spinach is not available, I use chard, kale, or occasionally bok choy.

Buckwheat Enchiladas with Black Beans and Chipotle Tomato Sauce

One of my favorite make-ahead meals is enchiladas. If a friend is having a baby, I'll often make up a couple of pans of enchiladas to tuck in her freezer. Even with a slightly longer list of ingredients, these enchiladas come together fairly quickly. Also, play around with the filling: mushrooms, sweet corn, tomatoes, and peppers all make lovely additions to the black beans and spinach.

YIELD: 8 ENCHILADAS

For the sauce:

1 tablespoon (15 ml) olive oil

½ medium (80 g) red onion, minced

1½ pounds (680 g) Roma tomatoes, coarsely chopped

½ to 1 teaspoon ground chipotle chile powder

½ teaspoon sea salt

2 teaspoons honey

½ cup (8 g) chopped fresh cilantro

For the crepe "tortillas":

½ cup (60 g) buckwheat flour

¼ teaspoon sea salt

2 large eggs

½ cup (118 ml) vegetable broth

2 tablespoons (30 ml) olive oil or melted coconut oil

For the filling:

1½ cups (60 to 70 g) loosely packed spinach, coarsely chopped

1 cup (180 g) cooked black beans, rinsed and drained if using canned

4 ounces (115 g) queso fresco, divided

¼ teaspoon sea salt

¼ cup (4 g) chopped fresh cilantro

1 tablespoon (15 ml) fresh lime juice

To make the sauce: Heat the olive oil in a saucepan over medium heat. Add the onion, and cook until translucent and fragrant, 6 to 8 minutes. Add the tomatoes, then stir in the chile powder, salt, and honey. Continue to cook until tomatoes break down, 10 to 15 minutes. Stir in the cilantro and remove from the heat. Using an immersion or regular blender, purée the sauce until smooth. Set aside.

To make the crepe "tortillas": Combine the buckwheat flour, salt, eggs, broth, and oil in a bowl, whisking until smooth. Heat an 8-inch (20-cm) skillet over medium heat and lightly grease with oil. Pour a scant ¼ cup (40 g) batter into the pan, and tilt so that the batter covers the entire pan. Cook for about 30 seconds. The crepe should look set and pull away from the edge of the pan. Flip and cook for another 15 seconds. As the crepes finish, layer on a plate, barely overlapping.

Preheat the oven to 375°F (190°C, or gas mark 5).

To make the filling: In a medium bowl, toss together the spinach, black beans, half the queso fresco, the salt, cilantro, and lime juice.

To assemble the enchiladas, pour half of the tomato sauce in the bottom of a 9 × 9-inch (23 × 23-cm) baking dish. Place 1 tortilla on a work surface and spread ⅓ cup (47 g) filling around the center. Roll tightly and place seam side down in the dish. Continue with the remaining tortillas and filling. Cover the enchiladas with the remaining tomato sauce and sprinkle the remaining 2 ounces (55 g) queso fresco over the top. Bake for 25 to 30 minutes, until the cheese is slightly brown.

SORGHUM

Growing up, I always saw three crops when I traveled through the vast farm belt in the United States: corn, soybeans, and sorghum. As a child, I understood that the majority of the plants growing were used for livestock feed and ethanol, and the corn and soybeans I ate were different. As for sorghum, I assumed the plant was grown only for feeding livestock. I was wrong, however, and little did I know that I was missing out on a wonderful grain.

As a staple cereal grain, sorghum is grown around the world for various uses such as livestock feed, alcohol, and even a sweet syrup that is similar to molasses. Thought to have originated in Egypt 8,000 years ago, sorghum is grown in Africa and Asia for human consumption, while the majority of the sorghum grown in the United States is used for animal feed and ethanol production. There are many different varieties of sorghum, ranging from a lighter, almost white color to a dark brown. The lighter-colored sorghum has a neutral flavor, while the darker has a more pronounced, earthy flavor. Sorghum is unique in that the outer hull is edible, keeping nearly all the nutrients in the grain intact. The outer shell of sorghum locks in moisture in a way similar to that of corn, so in addition to eating the grain and grinding the flour, sorghum can be popped for a great snack.

In the United States, sorghum can be hard to find. Some health food stores may sell it, but I have better luck ordering sorghum from online retailers.

Sorghum Flour

Because sorghum has its shell, it is best ground in a grain mill, high-powered blender, or coffee mill for small batches. A food processor or regular blender will most likely not be able to crack the tough exterior.

Sorghum flour is one of my favorite flours for gluten-free baking. White sorghum makes for a neutral flour that doesn't come with some of the grittiness associated with other gluten-free flours. Although sorghum flour can be used alone in recipes, I find the flour really shines when combined with other gluten-free flours, especially oat and buckwheat. If you are looking to create an all-purpose gluten-free mix, I recommend including sorghum flour.

Weights and Measurements

1 cup sorghum = 204 g

1 cup sorghum flour = 136 g

1 cup (204 g) sorghum = 1½ cups (204 g) sorghum flour

CINNAMON COFFEE CAKE

One of my favorite thank-you gifts to give to people is a fresh loaf
of quick bread. This coffee cake does not disappoint. The texture is moist and crumbly,
which complements the warm cinnamon flavor. The first time I made this bread,
my husband couldn't believe it was gluten-free.

YIELD: ONE 8-INCH (20-CM) LOAF

½ cup (68 g) sorghum flour

¼ cup (25 g) oat flour

¼ cup (36 g) brown rice flour

¼ cup (32 g) tapioca starch

1 teaspoon baking powder

¼ teaspoon baking soda

¼ teaspoon sea salt

½ cup (112 g) butter, melted and slightly
    cooled

⅓ cup (80 ml) maple syrup

2 large eggs

½ cup (112 g) whole-milk plain Greek
    yogurt

1 teaspoon vanilla extract

¼ cup (60 g) packed brown sugar

¼ cup (25 g) pecan halves, chopped

2 teaspoons ground cinnamon

Preheat the oven to 350°F (180°C, or gas mark 4). Lightly butter a
5 × 8-inch (13 × 20-cm) bread pan.

In a medium bowl, stir together the flours, tapioca starch, baking powder,
baking soda, and salt. In a separate bowl, whisk together the melted
butter, maple syrup, eggs, yogurt, and vanilla. Pour the wet ingredients
into the dry ingredients and stir until combined.

In a small bowl, mix together the brown sugar, pecans, and cinnamon.
Pour half the coffee cake batter into the prepared pan and sprinkle in
three-quarters of the brown sugar mixture. Cover the mixture with remaining batter and sprinkle the remaining brown sugar mixture on top.

Bake for 45 to 50 minutes, until the coffee cake is golden brown and
a knife inserted comes out clean. Let cool slightly in the pan before
transferring to a wire rack to cool completely. Store the coffee cake in
an airtight container at room temperature for 2 to 3 days.

Chocolate Espresso Doughnuts

My all-time favorite doughnut was a simple chocolate cake
doughnut smothered with chocolate icing from the local doughnut shop in
my hometown. When the shop closed, I set out to start making doughnuts at home, and
though these doughnuts aren't exactly the same as my memory of that chocolate
doughnut, they are equally—if not slightly more—delicious.

YIELD: 12 DOUGHNUTS

For the doughnuts:

1 cup (136 g) sorghum flour

½ cup (50 g) oat flour

⅔ cup (133 g) organic cane sugar

2 tablespoons (16 g) arrowroot

2 tablespoons (16 g) cornstarch

¼ cup (20 g) unsweetened cocoa powder

2 tablespoons espresso powder

2 teaspoons baking powder

¼ teaspoon sea salt

¼ cup (56 g) butter

¼ cup (56 g) semisweet chocolate chips

4 large eggs

¼ cup (56 g) whole-milk plain Greek
  yogurt

¼ cup (60 ml) 2% or whole milk

For the icing:

¾ cup plus 2 tablespoons (196 g)
  semisweet chocolate chips

¼ cup plus 2 tablespoons (90 ml)
  heavy cream

Sea salt, for garnish

Preheat the oven to 350°F and lightly grease a 12-well doughnut pan with butter.

To make the doughnuts: Combine the flours, sugar, arrowroot, cornstarch, cocoa powder, espresso powder, baking powder, and salt in a large bowl. In a double boiler or a heat-proof bowl set over a pot of simmering water, melt the butter and chocolate chips together. Remove from the heat and let cool slightly. In a separate bowl, whisk together the eggs, yogurt, and milk. Pour the butter mixture and the egg mixture into the dry ingredients. Stir until the batter is combined.

Spoon the batter into the doughnut pan, filling each doughnut well almost all the way full. Bake for 12 to 15 minutes, until the doughnuts spring back when pressed. Remove from the pan and place on a wire rack while you make the icing.

To make the icing: Combine the chocolate chips and cream in a double boiler or heat-proof bowl set over a pot of simmering water. Stir until the chocolate is melted. Dip the top half of a doughnut into the chocolate and return to the cooling rack. Let the icing dry slightly before sprinkling with salt. Store the doughnuts in an airtight container at room temperature for 2 to 3 days.

RICOTTA PANCAKES
WITH FRESH BERRIES

The first time I made ricotta pancakes, I was fairly skeptical
about the ricotta and the egg whites making a big difference. However,
I was delightfully surprised and inspired by how light and fluffy the pancakes
turned out. They've since become a weekend treat that pairs
especially well with fresh summer berries.

YIELD: 10 TO 12 PANCAKES

¾ cup (180 g) whole-milk ricotta

3 large eggs, separated

½ cup (120 ml) 2% or whole milk

1 tablespoon (15 ml) maple syrup

½ cup (68 g) sorghum flour

¼ cup (30 g) millet flour

2 tablespoons (16 g) tapioca starch

½ teaspoon baking powder

¼ teaspoon sea salt

Zest from ½ lemon

Fresh berries, for serving

Place the ricotta in a square of cheesecloth and squeeze out the excess
liquid. Combine the drained ricotta in a bowl with the egg yolks, milk,
and maple syrup. In a separate bowl, stir together the sorghum flour,
millet flour, tapioca starch, baking powder, salt, and lemon zest. Pour
the wet ingredients into the dry ingredients and stir until just combined.

Using a hand mixer or a stand mixer fitted with the whip attachment,
beat the egg whites until stiff. Spoon one-quarter of the egg whites into
the pancake batter and stir until incorporated to lighten up the batter.
Fold in the remaining egg whites.

Heat a griddle or skillet over medium-low heat and grease with butter.
For each pancake, pour ¼ cup (60 g) of the batter onto the griddle and
cook on each side until golden, about 2 minutes per side. Serve with
fresh berries.

You may be tempted to use part-skim ricotta, but I highly recommend sticking
with the full-fat variety. Although skim milk will work, the pancakes will lack the
lovely fluffy texture.

SWEET RICE

Straight out of college, I had dreams of eventually opening my own bakery and so found work at a popular local bakery to gain experience. The owner had just remodeled part of the building into a gluten-free kitchen, and luckily, I worked in there part-time. Working in the gluten-free kitchen exposed me to many of the flours in this book, especially in terms of baking. On days we would have a big order of cookies, I would make a run to the local Asian market and buy the store out of its stock of sweet rice flour. At the time, I didn't quite understand why this flour was so important.

Sweet rice is also labeled as glutinous rice, named for the rice's sticky qualities. Do not be confused by either name: Sweet glutinous rice contains no gluten and is not sweet. This rice is grown and primarily used in Eastern Asian cultures but is also grown in the United States. When I first started working with sweet rice at the bakery, we used sweet white rice. However, as I started learning more about whole grains and grinding flour at home, I started purchasing sweet brown rice. The difference between sweet brown rice and sweet white rice is similar to that between regular brown and white rice. Sweet brown rice is considered a whole grain, with nutrients still intact, while sweet white rice is processed, with the nutrients stripped away.

Sweet brown rice can be found at your local health food store or through a few online stores. Look for *sweet* rice and not just short-grain rice. There is a difference that will affect your cooking and baking. Sweet rice is also good to have on hand to make sticky rice.

Sweet Rice Flour

Sweet rice flour is slightly magical in terms of gluten-free baking. It acts as a binding agent for baked goods because of its high starch content (which is why the rice is sticky when cooked). Sweet rice flour is also used a lot in baking gluten-free breads, as it helps bring an elasticity that is often found only in gluten flours, primarily wheat flours.

I've found that sweet brown rice flour works just as well as sweet white rice flour. The taste of the flour is subtle and isn't intrusive to the overall flavor of the dish. Sweet rice flour is often included in gluten-free mixes but can be used on its own, as in the Chocolate Mochi Cake on page 129.

Grinding sweet rice flour is best done in a grain mill or high-powdered blender. Part of sweet rice flour's appeal is that the flour is fine and doesn't have a gritty texture in baked goods. Sweet rice can be ground in a coffee grinder, but sift the flour before using. I often grind a large batch of sweet brown rice into flour and store it in the freezer for use in sauces and gluten-free baking.

Weights and Measurements

1 cup sweet brown rice = 190 g

1 cup sweet brown rice flour = 152 g

1 cup (190 g) sweet brown rice = 1¼ cups (190 g) sweet brown rice flour

CHOCOLATE MOCHI CAKE

When it comes to food, texture is about as important as taste.
I have had food that I thought tasted decent . . . only to be turned off by an odd texture.
The first time I tried mochi cake, the dense, sponge-like texture slightly confused me.
After my initial reaction, however, I fell in love. This chocolate version is a favorite of
mine. The baking soda is optional, but I prefer the slightly lighter result it delivers.
Without the baking soda, the cake is extremely dense.

YIELD: 16 SERVINGS

¼ cup (56 g) butter
½ cup (112 g) dark chocolate chips
1 cup (152 g) sweet brown rice flour
½ cup (100 g) organic cane sugar
1 teaspoon baking soda
1 cup (226 g) whole-milk plain
   Greek yogurt
½ cup (120 ml) cup whole milk
1 large egg
1 teaspoon vanilla extract

Preheat the oven to 350°F (180°C, or gas mark 4). Lightly grease a
9 × 9-inch (23 × 23-cm) baking pan with butter.

In a double boiler or heat-proof bowl set over a pot of simmering water,
combine the butter and chocolate chips. Stir occasionally until melted
and smooth. Remove from the heat and let cool slightly.

In a large bowl, stir together the sweet brown rice flour, sugar, and baking soda. In a separate bowl, whisk together the yogurt, milk, egg, and
vanilla. Add the yogurt mixture and the slightly cooled chocolate mixture
to the flour mixture. Stir until well combined.

Pour the batter into the prepared pan. Bake until the cake no longer
jiggles, 25 to 28 minutes. Let cool before slicing. Store in an airtight
container at room temperature for 2 to 3 days.

ANGEL FOOD CAKE
WITH RICE FLOUR

One of my earliest memories of watching my mom bake involves
angel food cake. Never one to be patient, I'd stand at the edge of the kitchen and
watch as she flipped the baked cake onto an old glass pop bottle we kept for this specific
purpose. As the cake cooled, I would stare, half expecting the cake to fall, but it never
did. Once it cooled, we would dive into the cake with a bit of vanilla ice cream and
fresh fruit—a perfect antidote for the hot summers we had in Illinois.

YIELD: 8 TO 10 SERVINGS

1½ cups (320 g) egg whites (from
    10 or 11 eggs)
1½ teaspoons cream of tartar
1 cup (200 g) organic cane sugar
1 teaspoon vanilla extract
⅓ cup (50 g) sweet brown rice flour
¼ cup (36 g) brown rice flour
¼ cup (32 g) cornstarch
2 tablespoons (16 g) tapioca starch
¾ teaspoon sea salt

Preheat the oven to 350°F (180°C, or gas mark 4).

In the bowl of a stand mixer fitted with the whip attachment, whip the
egg whites on medium-high speed until thickened, 3 to 5 minutes. Add
the cream of tartar and continue to whip and thicken the egg whites.
Next, add the sugar in a slow, steady stream, letting it fully incorporate.
Continue to whip the mixture until the egg whites hold stiff peaks and
the batter has a glossy sheen to it, 3 to 5 more minutes. Turn the mixer
down to low and add the vanilla, mixing until just incorporated. Remove
the bowl from mixer.

In a separate bowl, combine the flours, starches, and salt. Scoop the
mixture into a sifter and sift one-quarter of it over the egg whites. Fold
the mixture into the egg whites, being careful to not overwork the egg
whites. Continue with the remaining flour mixture.

Pour the batter into a 10-inch (25.5-cm) tube pan and bake for 40 to
50 minutes, until golden across the top. Invert the cake pan onto the
neck of a glass bottle. Let cool completely before removing the cake
from pan.

Egg whites can be temperamental. Make sure the mixing bowl is clean and that
the egg whites do not have any trace of yolk. To ensure this, I crack 1 egg at a
time into a small bowl, separate the yolk from the egg white, and then place the
egg white in the mixing bowl.

ORANGE POPPY SEED CAKE

One of my favorite parts of planning and preparing for a road trip
is coming up with the snacks. I always pack a container of fruit, a bag of nuts, some
cheese, and some kind of fresh baked good. This cake is a perfect way to start a road trip
morning. The orange flavor is subtle and can easily be swapped for lemon,
creating my other favorite kind of poppy seed cake.

YIELD: 10 TO 12 SERVINGS

For the cake:

1 cup (144 g) sweet brown rice flour

½ cup (68 g) sorghum flour

½ cup (50 g) oat flour

¼ cup (32 g) arrowroot

2 tablespoons (12 g) poppy seeds

2 teaspoons baking powder

¼ teaspoon sea salt

½ cup (112 g) butter, melted

½ cup (120 ml) maple syrup

3 large eggs

2 tablespoons (12 g) orange zest

2 teaspoons vanilla extract

For the glaze:

2 tablespoons (26 g) organic cane sugar

2 tablespoons (30 ml) fresh orange juice

Preheat the oven to 350°F (180°C, or gas mark 4). Grease a 5 × 8-inch (13 × 20-cm) bread pan with butter.

To make the cake: In a large bowl, combine the flours, arrowroot, poppy seeds, baking powder, and salt. In a separate bowl, whisk together the melted butter, maple syrup, eggs, orange zest, and vanilla. Pour the wet ingredients into the dry ingredients and stir until the batter is smooth.

Pour into the prepared pan and bake for 40 to 45 minutes, until the top has domed and a knife comes out clean when inserted. Let cool in the pan for 10 minutes. Once the cake has cooled slightly, transfer from the pan to a wire rack.

To make the glaze: Whisk together the sugar and orange juice. Pour evenly over the top of the cake. Continue to let the cake cool before cutting. Store in an airtight container at room temperature for 2 to 3 days.

CHAPTER 4

Lovely Legumes

Some flours have a wide array of uses in my kitchen; bean flours, however, serve very specific purposes. Chickpea flour is always used to make socca, like in the version on page 142. Black bean, lentil, and split pea flours make satisfying, thick soups to serve during the winter months, and white bean flour makes a terrific thickening agent. For this book, I've stepped out of my usual comfort zone for bean flour to give you a few extra ideas that I hope will spur you to try these flours. This chapter includes a sampling of the many varieties of beans out there, and I recommend experimenting by grinding the ones you have on hand.

The main complaint that I, along with many others, have with bean flours is flavor. Bean flours make for wonderful gluten-free substitutions but can sometimes leave a less-than-pleasant bean-y or grassy aftertaste. Quite a few times I've been excited to try a gluten-free baked good only to have my hopes dashed because it tasted like I was eating a dried bean.

To mellow the bean flavor, I cook the beans first. While it is an extra step, cooking the beans and then dehydrating them also helps aid in digestion. I've found that eating bean flour that hasn't been precooked gave me uncomfortable gas

pains. When I prepare dried beans to eat, I soak and cook an extra-large batch. Half the batch goes into the freezer while the other half goes into the dehydrator until they are once again dried. The only exceptions to this rule are split peas and lentils, as both are have a milder flavor than beans and also seem to be a bit easier to digest.

If the taste of bean flour is still a bit strong for you, try toasting the bean flour before use. Add about 1 cup (120 g) of bean flour to a skillet over medium-low heat, and whisk continuously until lightly toasted, 8 to 10 minutes. I find this especially helpful with strongly flavored flours such as fava bean flour.

You may notice that many of my bean flour recipes are for soups and sauces. I enjoy using bean flours as thickening agents instead of milk because they add a bit of creaminess and a small amount of protein. My favorite recipe for using bean flour as a thickener is the Cheesy Fava Bean Dip on page 156. Bean flours can also be used in combination with gluten-free flours in baked goods. This is a great way to add extra protein and nutrients. I recommend starting with white bean or chickpea flour, as they both have a mild taste.

HOW I COOK BEANS

I use quite a few beans in my kitchen, and about every other week I find myself cooking dried beans. Portion out 3 to 4 cups (540 g to 720 g) of dried beans and sort through them to make sure there is no debris. Rinse the beans and then place them in a large pot with enough water to cover them by 2 or 3 (5 to 7.5 cm) inches. Soak overnight, or for 8 to 10 hours if soaking during the day. Drain and rinse the beans again, and return them to the pot along with fresh water to cover. Bring to a boil and cook for 40 to 45 minutes, until the beans are tender.

HOW I FREEZE BEANS

Because I cook large batches to use throughout the week, freezing them is a great way to keep the cooked beans from going bad. To freeze, I divide the beans and their cooking liquid into jars and let cool to room temperature. I use freezer-safe pint jars or freezer bags. For the most part, I freeze the beans in their liquid just in case I need it for a recipe, and I feel it helps prevent freezer burn if I store the beans longer than 2 weeks. If I know I'll want beans for salads, I freeze a couple of jars without the liquid that I can easily grab and use. To thaw, simply remove the beans ahead of time and place in the refrigerator, or place the container in a bowl filled with warm water to thaw faster.

CHICKPEAS
(GARBANZO BEANS)

When asked what my favorite ingredient is in the kitchen, I almost always respond with chickpeas. Not only are these beans packed full of protein, but they are also easily thrown into salads, made into spreads, and used as flour. For cooking, I soak and boil a big batch of dried beans and then freeze them to use in salads, soups, and hummus. Cooking and freezing large batches of dried beans is a great way to always have a meal or snack on hand, without having to reach for a can of beans.

Chickpeas, also known as garbanzo beans, come in two distinct varieties: kabuli and desi. Kabuli chickpeas are lighter in color, larger, and more typically sold in the United States, while desi chickpeas are smaller, darker, and more popular worldwide. These beans have a rich history, dating back 3,000 years to the Middle East. Chickpeas, both as whole beans and flour, are a great source of protein, zinc, and folate.

Chickpeas can be found in supermarkets and health food stores, both in dried and cooked forms. If purchasing dried beans from a bulk bin, check over the beans before cooking or grinding into flour. Rocks and odd-colored chickpeas may be found and should not be used.

Chickpea Flour

Of all the bean flours, chickpea flour, used extensively in Middle Eastern cooking, has gained popularity in the United States as a gluten-free alternative flour. Chickpea flour makes a great addition to any meal.

I should note, however, that chickpea flour can have a bean flavor. I've found that even after attempts at masking the flavor in recipes, I still taste beans. As such, the recipes I've included here are meant to work *with* the flavor rather than hide it. If the flavor doesn't bother you, trying substituting chickpea flour in baked good recipes as one of the gluten-free flours.

Because I live in the United States, I primarily grind kabuli chickpeas, resulting in a lighter, slightly cream-colored flour. Depending on the grain mill, chickpeas can be a little tough to grind because of their large size and odd shape. High-speed blenders and coffee grinders work great, as long as the flour is sifted afterward. If cooking and dehydrating the beans before grinding, make sure the beans are fully dried first, because soft beans can clog your grinder. During the cooking and dehydrating process, skins may loosen. I discard the separated ones but leave the intact skins on the beans.

Weights and Measurements

1 cup chickpeas = 180 g

1 cup chickpea flour =120 g

1 cup (180 g) chickpeas = 1½ cups (180 g) chickpea flour

BAKED CHICKPEA
FALAFEL SALAD

I've always had a weakness for certain fried foods, but
when I started revamping my diet to eat healthier, many of those foods became
ones that didn't taste good to me anymore. However, I am always up for a good crisp
falafel. This baked falafel is the perfect substitute for its fried counterpart.

YIELD: 4 SERVINGS

For the falafel:

1 clove garlic, peeled

½ cup (80 g) onion, coarsely chopped

2 cups (320 g) cooked chickpeas, rinsed
   and drained if using canned

½ cup (6 g) loosely packed fresh cilantro

¼ cup (15 g) fresh parsley

1 teaspoon each ground coriander and
   ground cumin

½ teaspoon paprika

¼ teaspoon sea salt

½ teaspoon baking powder

2 tablespoons (30 ml) each olive oil

fresh lime juice

¼ cup (30 g) chickpea flour

For the salad:

3 to 4 handfuls lettuce

1 cup cherry tomatoes

½ medium cucumber, sliced

3 to 4 ounces (85 to 115 g) feta, crumbled

For the dressing:

¼ cup (60 ml) each fresh lemon juice and
   extra-virgin olive oil

1 tablespoon (20 g) honey

1 garlic clove, minced

¼ teaspoon sea salt

Preheat the oven to 425°F (220°C, or gas mark 7). Lightly oil a baking
sheet or line it with parchment paper.

To make the falafel: Pulse the garlic clove in a food processor until
minced. Add the onion and continue to pulse with the garlic until
minced. Add the remaining falafel ingredients and run the food processor
until the chickpeas break down into small pieces and the mixture comes
together and is not too sticky. Shape 12 patties ½ inch (1.2 cm) thick
using 2 to 3 tablespoons (41 g) of the mixture per patty. Place on the
prepared baking sheet.

Bake for 20 minutes, flip, and bake for another 15 to 20 minutes, until
the patties are golden and crispy.

To make the salad: Assemble the lettuce, tomatoes, cucumbers, and feta
into a salad on a platter.

To make the dressing: In a jar with a lid, shake together the lemon juice,
olive oil, honey, garlic, and salt. Taste and adjust the flavors to your
liking. Lay the falafel patties on top of the salad. Pour the dressing
over the salad.

The falafel mixture can be made ahead of time and stored in an airtight container
in the refrigerator.

TOMATO BASIL SOCCA PIZZA

I hesitate to call this dish a pizza because the "crust" is a softer base
that makes it perfect for eating with a knife and fork. However, every time I make this,
it feels more and more like pizza. Try adding a few different cheeses to the rotation
to mix up the flavors. I love goat cheese, Gouda, and fontina.

YIELD: ONE 10-INCH (25.5-CM) PIZZA

1 cup (120 g) chickpea flour
1 cup (235 ml) water
¼ cup (60 ml) olive oil, divided
1 clove garlic, minced
¼ teaspoon sea salt
1 large (220 g) tomato, sliced ¼ inch
  (6 mm) thick
1½ cups (3 ounces) shredded mozzarella
3 or 4 basil leaves, julienned

In a bowl, whisk together the chickpea flour, water, 2 tablespoons
(30 ml) of the olive oil, the garlic, and salt. Let sit for 1 hour.

Turn on the broiler with a rack positioned 8 inches (20 cm) from the
heat and a place a 10-inch (25.5-cm) ovenproof skillet in the oven to
preheat. Once the skillet is hot, carefully remove from the oven and add
1 tablespoon (15 ml) olive oil. Swirl around to cover the bottom. Pour in
the chickpea batter and return the skillet to the broiler. Cook for 5 to
8 minutes, until the socca is set and the edges are browning. Remove
from oven, turn off broiler, and turn the oven to 425°F (220°C, or
gas mark 7).

Spread the remaining 1 tablespoon (15 ml) olive oil on top of the socca.
Layer the tomato slices around the socca. Sprinkle the cheese on top and
return the skillet to the oven. Bake for 8 to 10 minutes, until the cheese
is browning and the socca is crisp. Remove from the oven and sprinkle
the basil on top. Let cool for 2 to 3 minutes before serving.

Basic Herbed Socca with Sun-Dried Tomato Spread

I would feel remiss if, in a chapter about bean flours, I didn't include a recipe for socca. This pancake-like bread is traditionally made from chickpea flour. The first time I had socca was also the first time I realized that beans could be made into a useful flour. The flavor of the chickpeas plays well with the herbs and the sun-dried tomato spread. Feel free to play around with the herbs. I often toss in whatever herbs I have on hand at the time.

YIELD: ONE 10-INCH (25.5-CM) FLATBREAD AND 1 CUP (295 G) SPREAD

For the socca:

1 cup (120 g) chickpea flour

1 cup (235 ml) water

3 tablespoons (45 ml) olive oil, divided, plus more for drizzling

1 clove garlic, minced

¼ teaspoon sea salt

2 teaspoons fresh thyme

2 teaspoons chopped fresh parsley

For the spread:

¼ cup (32 g) sun-dried tomatoes (dry-packed)

½ cup (118 ml) water

⅓ cup (80 g) ricotta

¼ cup (25 g) walnuts

1 tablespoon (20 g) honey

1 tablespoon (15 ml) fresh lemon juice

2 tablespoons (5 g) fresh basil

To make the socca: One hour ahead, start the socca. In a mixing bowl, whisk together the chickpea flour, water, 2 tablespoons (30 ml) of the olive oil, the garlic, salt, thyme, and parsley. Let the batter sit for 1 hour.

To make the spread: While the batter sits, combine the sun-dried tomatoes with the water in a small bowl to rehydrate the tomatoes.

Turn on the broiler with a rack positioned 8 inches (20 cm) from the heat and place a 10-inch (25.5-cm) ovenproof skillet in the oven to preheat. Once the skillet is hot, carefully remove from the oven and add the remaining 1 tablespoon (15 ml) olive oil. Swirl around to cover the bottom. Pour in the chickpea batter and return the pan to the broiler. Cook for 8 to 10 minutes, until the outer edges of the socca are browning and the center is set.

To finish making the spread, drain the sun-dried tomatoes and combine with the ricotta, walnuts, honey, and lemon juice in a food processor. Pulse until mixed. Add the basil and pulse 3 or 4 more times to combine. Taste and adjust the flavors as desired.

Serve the warm bread with a smear of sun-dried tomato spread and a light drizzle of olive oil.

Roasted red peppers and roasted garlic make for wonderful substitutions for the sun-dried tomatoes.

LENTILS

When I made the switch to a vegetarian diet, I knew that I wanted to focus on substantial meals that would keep me full and satisfied. I turned to beans, which eventually led me to discover the wonders of lentils. I found the taste of lentils to have a mellower flavor that soaked up herbs and spices wonderfully. I routinely cooked up a large batch to keep in the refrigerator for tossing on salads or tacos for weeknight meals, and a hefty bowl of curried lentils found its way into my weekly rotation. I also found that lentils were easier to prepare on a whim compared with black beans or chickpeas, which do better with a good overnight soaking.

Thought to be one of the first foods cultivated, lentils are believed to have originated in Central Asia. While there are at least a dozen popular lentil varieties, I cook and grind flour from red, green, brown, and Puy lentils (a dark green lentil with speckles). Lentils grow in pods that contain 1 or 2 seeds. After harvesting, the lentils may or may not go through a process of removing the husk and splitting. The husks of red lentils, for example, are removed and the lentils are split before selling. This gives them their quick cooking time and is also the main reason red lentils do not hold their shape when cooked. On the other hand, green and brown lentils take longer to cook but retain their shape for use in salads and soups.

Lentils can be found in the supermarket and at the health food store, both prepackaged and in bulk bins. I keep different varieties of lentils around for different uses, such as red lentils for curries, green lentils for salads, and brown lentils for hearty dishes, such as the lentil and pumpkin flour "meat"balls on page 179. I use the different lentils fairly interchangeably when I grind flour, unless I'm looking for a certain color, as in the Sweet Potato and Red Lentil Gnocchi on page 146.

Lentil Flour

I think that of all the legume flours, lentil is the most underrated. The flour has a wide array of uses as a thickening and binding agent. Lentil flour does not seem to have the overpowering bean flavor that comes from other legumes, which makes it great for adding to soups as a neutral thickener.

For grinding, I tend to stick with red or brown lentils, depending on the dish. If I want the vibrant red color, I use red lentils, but if I want a more traditional flour color, I grind brown lentils. Lentils are best ground in a grain mill or high-powdered blender but can also be ground in a coffee grinder. Lentils grind into extremely fine flour. Use caution when grinding or transferring the flour, as it has a tendency to go everywhere. I left my grinder once for 10 seconds and turned around to see a cloud of lentil flour blowing out of the canister.

Weights and Measurements

1 cup lentils = 200 g

1 cup lentil flour =140 g

1 cup (200 g) lentils = scant 1½ cups (200 g) lentil flour

FETA AND RED LENTIL DIP

I am notorious for never wanting to show up to a party
empty-handed, and of course, my go-to is always food. This dip is
easy to throw together and full of flavor. Serve with crackers,
toasted bread, or even sliced vegetables.

YIELD: 3 CUPS

2 cups (470 ml) water

½ cup (70 g) red lentil flour

1 clove garlic, peeled

8 ounces (225 g) feta, divided

1 teaspoon ground turmeric

2 teaspoons ground cumin

2 teaspoons ground coriander

½ teaspoon ground ginger

½ teaspoon crushed red pepper

1 tablespoon (15 ml) fresh lemon juice

2 tablespoons (30 ml) olive oil, plus
more for drizzling

In a medium saucepan, bring the water to a simmer and whisk in the
lentil flour. Continue to whisk until mixture thickens and is smooth.
Remove from the heat and let cool for 10 minutes.

In a food processor, pulse the garlic until minced. Add 6 ounces
(170 g) of the feta and pulse again until the feta is broken down into
small pieces. Add the lentil mixture, turmeric, cumin, coriander, ginger,
crushed red pepper, lemon juice, and olive oil. Turn the processor on and
let it run until the mixture comes together, adding more olive oil if the
texture is too thick.

Scoop the dip into the saucepan over medium-low heat. Whisk, adding
more water if you desire a thinner texture. Serve the dip warm with the
remaining 2 ounces (55 g) feta crumbled on top and a drizzle of olive oil.

Quick Lentil Curry Soup

Leftover brown rice is a staple in my refrigerator, and most
of the time I just sauté a few veggies to eat with it. However, sometimes I fix this
quick soup to smother the brown rice for a rich and satisfying meal that comes together
in less than 20 minutes. Find a good curry spice blend from a spice shop or
Indian grocer to always have on hand for meals like this.

Yield: 2 servings

2¼ cups (530 ml) vegetable broth

¼ cup (35 g) red lentil flour

2 teaspoons curry powder

¼ cup (56 g/60 ml) plain Greek yogurt or
canned coconut milk

1 to 2 tablespoons (15 to 30 ml) fresh
lime juice

¼ cup (4 g) fresh cilantro

In a large saucepan, whisk together the vegetable broth, lentil flour,
and curry powder. Bring to a boil and then reduce the heat to a simmer.
Cook, whisking frequently, for 10 minutes.

Remove from the heat and stir in the yogurt or coconut milk. Serve with a
squirt of lime juice and a sprinkle of cilantro.

Depending on the curry powder, 2 teaspoons may not be enough. Taste and add
more curry powder if the flavor is not strong enough for you.

SWEET POTATO AND RED LENTIL GNOCCHI WITH PESTO CREAM SAUCE

I am always on the lookout for delicious, hearty meals that will help
fill me up during the winter months. While I always love a good bowl
of pasta, I find it's easy to go heavy on the carbs and ditch the protein.
Using lentil flour in gnocchi is a great way to have a bowl of
comfort food that also happens to have protein and nutrients.

YIELD: 4 SERVINGS

1 medium (230 g) sweet potato

For the sauce:

½ cup (20 g) packed fresh basil leaves

1 clove garlic, peeled

1 tablespoon (9 g) pine nuts, toasted

3 tablespoons (45 ml) olive oil

1 tablespoon (15 ml) fresh lemon juice

¼ cup (8 g) grated Parmesan, plus more
for serving

¼ teaspoon sea salt

⅔ cup (160 ml) heavy cream

For the gnocchi:

1 egg yolk

½ teaspoon sea salt

½ teaspoon freshly ground black pepper

1½ to 2 cups (210 to 280 g) red lentil
flour, divided

Preheat the oven to 425°F (220°C, or gas mark 7).

Pierce the sweet potato with a fork and bake for 30 minutes, or
until soft.

To make the sauce: In a food processor, combine the basil, garlic,
pine nuts, olive oil, lemon juice, Parmesan, and salt. Process until
well combined. It's okay if the pesto is thick. Transfer the pesto to
a large skillet, add the cream, and set aside.

To make the gnocchi: When cool enough to handle, remove the sweet
potato from the skin, mash, and measure ½ cup (112 g). Reserve the rest
for another use. Combine the sweet potato, egg yolk, salt, pepper, and ½
cup (70 g) of the red lentil flour. Continue to stir and add flour until the
mixture forms a soft dough that is still slightly sticky. Place on a surface
covered in red lentil flour. Divide the dough into 4 balls and roll each out
into 1-inch (2.5-cm)-thick ropes. With a bench scraper or knife, cut each
rope into 1-inch (1.2-cm) segments. Mark the gnocchi with the tines of a
fork.

Bring a pot of salted water to a boil and carefully add the gnocchi. Cook
for 4 to 5 minutes or until the gnocchi rise to the top. Remove from the
water with a slotted spoon and place in a bowl.

Over medium-low heat, heat the pesto mixture. Cook until it starts to
thicken, 4 to 5 minutes. Add the gnocchi and continue to cook for
1 to 2 more minutes. Remove and serve with Parmesan.

BLACK BEANS

Next to the jars of cooked chickpeas stashed in my freezer sit a similar amount of jars filled with black beans. Whenever I'm craving a quick and hearty meal, my go-to recipe is a simple black beans and rice dish. During my childhood, I went out of my way to avoid black beans, but I fell in love with them during a trip to Costa Rica, when my days would start with *gallo pinto*, or Costa Rican rice and beans. I always have dried and cooked black beans ready for a simple dinner, a delicious vegetarian taco, or a perfect black bean burger (page 149).

Black beans, sometimes called turtle beans, are characterized by a black, slightly shiny shell and a lighter inside. Black beans are a type of common bean that falls into the same family as pinto, navy, and kidney beans, and they are full of fiber and protein. Black beans have a mild, earthy flavor that pairs well with herbs and spices, especially cumin and coriander. Black beans originated in Central and South America and spread to Europe during the Spanish exploration of the fifteenth century, which is why they are used extensively in Latin cooking.

Black beans come canned or dried and can be found in supermarkets and health food stores. In addition to cooked beans, I keep a big jar of dried black beans in the house for cooking and grinding into flour.

Black Bean Flour

Black bean flour was a bit of a surprise to me the first few times I cooked with it. While many of the bean flours have a bit of a bitter aftertaste, I didn't always taste it with black bean flour. When I made pasta (see the recipe on page 150), I found myself eating the just-cooked pasta, waiting for the aftertaste to come, and it never did. Black beans' mild and slightly earthy flavor makes for a great flour that has a multitude of uses in baking, in pasta making, and as a binding agent. I've often seen black bean flour used in gluten-free brownie recipes.

Black beans work best when ground in a grain mill or high-powered blender and sifted before use. Sort through the beans before grinding, because occasionally rocks and debris can be found. This is especially important with black beans, as their color makes them hard to distinguish from a rock.

Weights and Measurements

1 cup black beans = 180 g

1 cup black bean flour = 120 g

1 cup (180 g) black beans = 1½ cups (180 g) black bean flour

BLACK BEAN AND SPINACH BURGERS

While I love the flavor of the Curried Sweet Potato and Teff Burgers on page 79, I occasionally crave a burger that blends a bit better with a wide variety of toppings. These burgers come together quickly and have a slight crunch due to the sunflower seeds.

YIELD: 4 TO 6 BURGERS

1 cup (180 g) cooked black beans, rinsed and drained if using canned

½ cup (80 g) diced onion

2 tablespoons (30 ml) olive oil

2 tablespoons (30 ml) fresh lime juice

1½ teaspoons smoked paprika

1 large egg

½ cup (72 g) sunflower seeds (raw or roasted)

¾ cup (90 g) black bean flour

1 teaspoon sea salt

2 cups (80 g) packed spinach

Preheat the oven to 375°F (190°C, or gas mark 5).

In the bowl of a food processor, combine the black beans, onion, olive oil, lime juice, paprika, egg, sunflower seeds, black bean flour, and salt. Pulse 4 or 5 times to break down the beans and blend the ingredients. Add the spinach and pulse until spinach is incorporated and the mixture is fairly well mixed together.

With wet hands, form 4 to 6 patties, place on a baking sheet, and bake for 10 minutes to set. On a slightly oiled grill or grill pan over medium heat, finish cooking the black bean patties until brown and crispy, 3 to 4 minutes per side. If not using right away, bake and freeze the patties, separating with waxed paper. Patties can be grilled directly from the freezer, without thawing.

Make sure the bean mixture is well blended before adding the spinach. If the food processor runs for too long with the spinach, your burgers will turn green.

BLACK BEAN BOW TIES WITH SWEET CORN HASH

This is one of my favorite pasta salads—nontraditional and
filled with Southwestern ingredients. This recipe pulls inspiration from
the classic pasta salad while merging the pasta and black beans
into one. This dough can be sticky; do not be afraid to use
a lot of black bean flour when rolling it out.

YIELD: 4 SERVINGS

For the dough:

2 cups (240 g) black bean flour

2 large eggs

3 tablespoons (45 ml) water

½ teaspoon sea salt

For the hash:

1 tablespoon (15 ml) olive oil

1 medium (160 g) red onion, diced

1 small jalapeño chile pepper, diced
(retain seeds for more heat)

Kernels from 2 large ears corn, or 2 cups
(326 g) frozen corn kernels, thawed

½ cup (8 g) minced fresh cilantro, plus
more for garnish

3 tablespoons (45 ml) fresh lime juice

1 tablespoon (20 g) honey

4 ounces (115 g) queso fresco, crumbled

1 lime, cut into wedges

To make the dough: On a clean surface, create a pile of black bean flour
and make a well in the center. Crack the eggs in the center, and add the
water and salt, lightly mixing together with a fork. Slowly incorporate the
flour into the egg mixture. Eventually, ditch the fork and work the dough
with your hands, adding more flour to the surface if needed to prevent
sticking. Once the dough is soft and not sticky, set aside and let it rest
for 10 minutes. Bring a pot of salted water to a boil and make the hash.

To make the hash: In a large skillet, heat the olive oil over medium-low
heat. Add the onion and jalapeño and cook for 6 to 8 minutes, until the
onion is translucent. Add the corn and cook for another 6 to 8 minutes,
until the corn is tender and browning. Stir in the cilantro, lime juice, and
honey. Cook for 1 minute longer.

Divide the pasta dough in half and place one half on a well-floured
surface. Roll the dough into a rectangle that is ¼ inch (6 mm) thick. With
a pizza cutter or knife, cut the pasta into 1 × 1½-inch (2.5 × 3.7-cm) rectangles and pinch the center of each rectangle along the longer egdges to
create a bow tie. Add to the boiling water in batches and cook for 4 to 5
minutes, until the pasta is tender and floating to the top.

Once the pasta and corn are cooked, toss them together in a large bowl
and serve with a sprinkle of queso fresco, more cilantro, and a squeeze of
lime juice.

BLACK BEAN SOUP TOPPED
WITH GREEK YOGURT

When eating at restaurants, I am always eager to hear about
the soup selection. I love soups because almost any combination of flavors
can meld to create a hearty one-bowl meal. Black bean soup always fascinates me
because on their own, black beans have a mild taste. Yet when paired with spices,
black beans shine and make any dish, especially soup, wonderful.

YIELD: 2 SERVINGS

1 tablespoon (15 ml) olive oil

½ cup (80 g) minced onion

1 clove garlic, minced

2 teaspoons ground cumin

1 teaspoon ground coriander

¼ cup plus 2 tablespoons (45 g) black
bean flour

3 cups (705 ml) vegetable broth

1 cup (180 g) cooked black beans, rinsed
and drained if using canned

½ cup (112 g) whole-milk plain Greek
yogurt

2 or 3 scallions, diced

In a large saucepan or stockpot, heat the olive oil over medium-low heat.
Add the onion and cook for 6 to 8 minutes, until soft. Stir in the garlic,
cumin, and coriander, and cook for 1 more minute. Add the black bean
flour and stir to coat the onions. Pour in the vegetable broth, mixing thoroughly with the black bean flour. Bring to a boil, then reduce the heat to
a simmer and add the black beans. Cook for 15 minutes, until thickened
and the taste of the black bean flour is gone.

Ladle into bowls and serve with a dollop of Greek yogurt and a sprinkle
of scallions.

FAVA BEANS

Out of all the beans in this book, fava beans were the first bean I ever ate fresh. I, being the always curious one, snagged a pound at the farmers' market because I had seen them in the bulk bins as dried beans but had never had a chance to try them fresh. Once I battled the pods to remove the actual beans, I ended up with something buttery and smooth. Whenever they are in season, I eat the beans fresh, but for all the other months of the year, I keep a small bag of dried beans.

Fava beans are a member of a species of bean called *broad beans*, characterized by their large size and flat shape. While fresh fava beans have a green color, dried favas range from tan to a yellowish green. Fava beans are native to Northern Africa and southern parts of Asia, and they've been widely used in the Mediterranean diet, as they were one of the first beans eaten in that region. Like other beans, fava beans are an excellent source of protein, dietary fiber, vitamins, and minerals. As with chickpeas, fava beans are great to keep around for a nice spin on hummus, an extra addition to salads, or even to serve on their own tossed in a bit of herb butter.

When purchasing dried fava beans, look for beans that have already been blanched and the seed coat removed. I have mixed reviews to offer on whether to cook the fava bean and dehydrate it before grinding. If left to soak and cook for too long, fava beans end up mushy, which makes them hard to dry. If you want to cook the beans before grinding, skip the soaking step, and cook for 30 to 40 minutes, just until the beans are starting to become tender. Fava beans can be found prepackaged or in bulk bins at health food stores.

Fava Bean Flour

The taste of fava bean flour is a bit stronger than that of other bean flours, and quite often, fava bean flour is mixed with chickpea flour to mitigate this. I use the flour sparingly in baked goods and primarily use it when I know another flavor will be equal or stronger. Also, fava bean flour has one of the least appealing smells after being ground. But don't let the smell dissuade you from using the flour—it makes a lovely addition to soups and dips.

The size of the fava bean can sometimes be a hindrance when trying to grind the dried bean. The opening for some grain mills is too small, or because of the bean size, the beans won't grind into fine powder. I've had the best luck with a high-powered blender or coffee grinder. It's best to work in small batches, sifting the beans a few times to separate the large pieces from the flour. However, be aware that the end result is a flour with a slightly gritty texture.

Weights and Measurements

1 cup fava beans = 180 g

1 cup fava bean flour =120 g

1 cup (180 g) fava beans = 1½ cups (180 g) fava bean flour

TOMATO AND FAVA BEAN SOUP

I have about ten different tomato soup recipes in my repertoire.
Among them is one brimming with heavy cream; another is made with
roasted tomatoes; and this one is made creamy by the addition of fava bean flour.
I often make a large batch of this soup during the peak of tomato season
to freeze and eat during the cooler fall weather.

YIELD: 3 OR 4 SERVINGS

2 tablespoons (30 ml) olive oil

2 cloves garlic, minced

2 pounds (910 g) Roma tomatoes, diced

½ cup (60 g) fava bean flour

4 cups (940 ml) vegetable broth

1 tablespoon (2.4 g) fresh thyme

1 tablespoon (4 g) fresh oregano

2 teaspoons organic cane sugar

½ teaspoon sea salt

½ teaspoon freshly ground black pepper

Grated Parmesan or shredded mozzarella,
  for serving (optional)

In a large saucepan or stockpot, heat the olive oil over medium-low heat. Add the garlic and cook for 1 minute. Stir in the tomatoes and cook until the tomatoes start to break down, 10 minutes.

Add the fava flour and cook for 1 minute. Add the vegetable broth, thyme, oregano, sugar, salt, and pepper. Continue to cook for 15 minutes. Remove from the heat and, with an immersion blender or regular blender, purée the soup until smooth.

Serve with a sprinkle of cheese, if desired.

CHEESY FAVA BEAN DIP

On almost a weekly basis, it seems that there is one night
when we do not feel like cooking. Our go-to meal for these nights is a fully
loaded plate of nachos: a few tortilla chips topped with veggies, avocados, salsa,
and of course, all smothered in nacho cheese. The fava bean flour creates
a creamy texture and melds perfectly with the cheese.

YIELD: ABOUT 1 CUP (332 G)

¼ cup (30 g) fava bean flour

1 to 1¼ cups (235 to 294 ml) water

1 small clove garlic, minced

½ cup (60 g) shredded white Cheddar

2 tablespoons (4 g) chopped fresh cilantro

½ teaspoon smoked paprika

¼ teaspoon sea salt

¼ teaspoon freshly ground black pepper

In a medium saucepan over medium-low heat, combine the fava bean flour, water, and garlic. Cook for 4 to 5 minutes to cook out the flour taste. Remove from the heat and whisk in the cheese and cilantro. If the dip is too thick, return to the heat and add 1 tablespoon (15 ml) water at a time until the texture thins to the desired consistency. Add the paprika, salt, and pepper and stir until well combined.

Carrot Ginger Soup

Soup is often a hearty food, but occasionally, I turn to this recipe
for a lighter soup. The carrots, ginger, and coconut milk come together
for a refreshing flavor. The fava bean flour adds a bit of heft
and creaminess to the soup.

YIELD: 2 SERVINGS

2 cups (240 g) ¼-inch (6-mm)-thick sliced
  carrots

½ medium (80 g) onion, sliced

1 tablespoon (15 ml) olive oil

2½ cups (588 ml) water

1 tablespoon (15 ml) tamari (soy sauce)

1 teaspoon minced fresh ginger

¼ cup (30 g) fava bean flour

¼ cup (60 ml) canned coconut milk

2 to 3 tablespoons (17 to 26 g) roasted
  pumpkin seeds

Preheat the oven to 400°F (200°C, or gas mark 6). Lightly grease a
baking sheet.

In a medium bowl, toss the carrots and onion with the olive oil. Spread in
a single layer on the prepared baking sheet. Roast for 20 to 25 minutes,
until the carrots are tender.

Combine the water, tamari, and ginger in a medium saucepan. Bring to a
boil, then reduce the heat to a simmer, and cook for 5 minutes. Whisk in
the fava bean flour and cook for 4 more minutes, stirring often.

Add the roasted carrots and onion and cook for 5 minutes, until the
carrots are very soft. Remove from the heat and, using an immersion
blender or regular blender, purée the soup. Strain if you desire a very
smooth soup.

Serve the soup with a swirl of coconut milk and a sprinkle of pumpkin
seeds.

SPLIT PEAS

I'm slightly ashamed to say that for quite a while, I would buy split peas from the bulk bins, bring them home, and then they would sit there, unused. I was always attracted to their bright color, which popped out from the tans and browns in the surrounding bulk bin containers. Eventually, I started experimenting by adding cooked peas to salads or using peas in place of lentils in some dishes. When I started grinding flours, I went straight to peas, primarily because I thought it would be fun to have yellow and green flours to use.

Dried split peas are exactly what the name implies: peas that have been dried, peeled, and split. The peeling process removes the outer pod while the splitting process removes the outer layer of the pea, leaving a brightly colored pea half. They come in the two common pea colors of yellow and green. The flavor of fresh peas and dried peas differs in that dried peas have an earthier, more concentrated flavor than the lighter, slightly sweeter flavor of fresh peas.

It's thought that dried peas originated in central Asia and Europe from a type of field pea and that dried peas were consumed before fresh peas. Peas are packed full of protein and fiber but are relatively low in fat, making them a perfect component of vegetarian cooking.

Split peas can be found in both traditional grocery stores and health food stores. Look for split peas with the other dried beans. I keep split peas on hand for creamy soups and curries, as they require no soaking and cook up quickly.

Split Pea Flour

The flavor of yellow and green split peas is subtle, and once the flour taste cooks out, I think the flour is barely noticeable. In recipes where the flour is used as a binding agent, do not get impatient and turn up the heat. The key to cooking out the flour taste is to cook low and slow (a concept that works with most flours but particularly with split peas). While I tend to cook with color, as you might notice in the recipes, the pea flours can be used interchangeably.

Split peas are best ground in a grain mill, a high-powered blender, or a coffee grinder. I always sift pea flour through a sieve before using to catch any small pieces that didn't grind into flour. Before grinding, sort through the peas to make sure any debris isn't present, such as small rocks or other varieties of beans.

Weights and Measurements

1 cup split peas = 216 g

1 cup split pea flour = 144 g

1 cup (216 g) split peas = 1½ cups (216 g) split pea flour

SPICED PEA SOUP

While whole green peas are traditionally used to make soup,
this variation, using pea flour, comes together quickly, and there is no need
to purée because the soup is already smooth. If you don't have a mortar
and pestle, use ground cumin and fenugreek—simply use a smidge
less to start with and increase as desired.

YIELD: 2 SERVINGS

1½ teaspoons cumin seeds
½ teaspoon fenugreek seeds
½ teaspoon ground turmeric
4½ cups (1058 ml) vegetable broth
½ cup (72 g) pea flour
Coconut milk, for garnish
Fresh lime juice, for garnish (optional)

In a dry skillet over low heat, toast the cumin and fenugreek seeds until fragrant, 4 to 5 minutes. Remove from the skillet and transfer to a mortar and pestle. Crush the spices and combine with the turmeric.

In a large saucepan, whisk together the spices, vegetable broth, and pea flour. Bring to a boil, then reduce the heat to a simmer and cook for 10 to 15 minutes, until the soup has thickened and the flour taste has cooked out. Ladle the soup into bowls and serve with a swirl of coconut milk and squeeze of lime juice, if desired.

Green Pea Flour and Zucchini Fritters with Dill Yogurt Dip

On my blog, I get many questions about what to do with produce,
and not surprisingly, zucchini ranks highest in those questions. It seems that during
the summer, if you turn your back for a day, zucchini grow tenfold, and after all, people
can only eat so much zucchini bread. These fritters are a great way to use up a mass
quantity of zucchini and make for a perfect light summer lunch.

YIELD: 16 TO 20 FRITTERS

For the fritters:

1 pound (about 455 g, or 2 medium) zucchini

1 tablespoon (6 g) lemon zest

1 tablespoon (15 ml) fresh lemon juice

1 clove garlic, minced

½ teaspoon sea salt

¼ teaspoon freshly ground black pepper

2 large eggs, lightly beaten

½ cup (72 g) green pea flour

½ cup (16 g) grated Parmesan

1 tablespoon (4 g) fresh dill

2 to 4 tablespoons (30 to 60 ml) olive oil

For the dip:

½ cup (112 g) whole-milk plain Greek yogurt

1 tablespoon (4 g) minced fresh dill

2 teaspoons honey

1 tablespoon (15 ml) fresh lemon juice

To make the fritters: Trim the ends off the zucchini and, using a box grater with large holes, grate the zucchini. Place in a bowl and combine with the lemon zest and juice, garlic, salt, and pepper. Pour the beaten eggs over the zucchini, along with the green pea flour, Parmesan, and dill. Stir until the zucchini is coated.

In a skillet with tall sides, heat 2 tablespoons (30 ml) olive oil over medium-low heat. Spoon the zucchini mixture into the skillet in 2- to 3-tablespoon (45 g) drops. Press down with the back of the spoon and cook on each side until crisp, 4 to 5 minutes. Transfer to paper towels to drain. Continue with the remaining zucchini mixture, adding 1 to 2 tablespoons (15 to 30 ml) more olive oil to the skillet as needed.

To make the dip: In a small bowl, whisk together the yogurt, dill, honey, and lemon juice. Serve with the hot and crispy fritters.

🖊 With the addition of green pea flour, it's important to cook these fritters over a lower temperature to allow the flour taste time to cook out. Do not get impatient and turn up the heat—the flavor will not be the same!

CORN CAKES WITH SPINACH CILANTRO SAUCE

Along with big bowls of salad, and fresh heirloom tomatoes served
with a sprinkle of sea salt, these corn cakes are a summer staple. I often whip up
a batch and cook half one night and half the next. If you are like me and prefer a little
heat, add a pinch or two of crushed red pepper to the corn cake batter.

YIELD: 2 OR 3 SERVINGS

For the corn cakes:

2 tablespoons (30 ml) olive oil, divided

¼ cup (40 g) minced red onion

Kernels from 2 medium ears corn, or
1¾ cups (300 g) frozen corn kernels,
thawed

¼ to ½ cup (36 to 72 g) yellow pea flour

¼ cup (4 g) chopped fresh cilantro

¼ teaspoon baking powder

1 egg yolk

1 tablespoon (20 g) honey

1 tablespoon (15 ml) fresh lime juice

¼ teaspoon sea salt

For the sauce:

½ cup (20 g) packed spinach

¼ cup (4 g) fresh cilantro

1 clove garlic, peeled

1 tablespoon (15 ml) olive oil

1 tablespoon (15 ml) fresh lime juice

1 tablespoon (15 ml) water

2 teaspoons honey

¼ teaspoon sea salt

To make the corn cakes: Heat 1 tablespoon (15 ml) of the olive oil in a large skillet over medium-low heat and add the onion. Cook for 6 to 7 minutes, until soft. Add the corn and cook for 6 minutes, until tender. Remove from the heat, place the mixture in a bowl, and wipe out the skillet. Add ¼ cup (36 g) of the pea flour, the cilantro, baking powder, egg yolk, honey, lime juice, and salt to the bowl; stir until well combined. Add more of the flour until the mixture pulls together and you can form patties that are slightly loose and wet.

Divide the batter into 6 mounds. Heat the remaining 1 tablespoon (15 ml) olive oil in the skillet over medium heat. Spoon the corn mounds into the skillet and pat in to 1-inch (2.5-cm)-thick disks. Fry on each side for 5 to 6 minutes, until browned.

To make the sauce: Combine the spinach, cilantro, garlic, olive oil, lime juice, water, honey, and salt in a food processor. Run until the sauce is smooth. Serve the corn cakes with the sauce.

WHITE BEANS

Before I started amassing a collection of bulk-bin grains and nuts, the only time I ever cooked or ate white beans was when I ate baked beans, and that was rare—I was not a baked bean eater. White beans never made a dent in my kitchen, primarily because I ate chickpeas and thought they fit well into any recipe. Turns out, I was slightly wrong. The smooth texture and neutral flavor of white beans works well in meals where beans need to meld with other flavors, such as herbs, spices, or even vegetables. White beans are on the short list of favorites, especially in salads, soups, and even on their own as a side dish.

The term *white bean* is generic and includes navy, Great Northern, and cannellini beans. Some white bean varieties, such as the navy bean, are from the line of common beans that originated in South America, specifically Peru. From there, those beans spread through South and Central America and were eventually brought to Europe by explorers. Other beans, such as the marrow bean, are thought to have originated in the Middle East. These varieties of white beans have similar nutritional value, all packed with protein and fiber. However, the beans differ in taste and size:

Navy beans: The smallest of the three, these beans have a mellow flavor. When cooked, these beans are a bit mushy and do not hold their shape as well as Great Northern or cannellini beans. Navy beans are the traditional bean used for baked beans.

Great Northern beans: This bean is smaller and when cooked, it has a slightly grittier texture than the cannellini bean but keeps its shape better than the navy bean. Great Northern beans also pack a bit of a nuttier flavor than navy or cannellini beans.

Cannellini beans: This large bean has a slight kidney shape that holds its form well when cooked. The texture is extremely smooth, and the bean has a mild, earthy taste.

Marrow beans: The largest of the white beans, marrow beans are known for their distinctive, strong flavor that is likened to the taste of meat. The texture of the cooked bean is creamy and not gritty. Marrow beans are often used as a whole bean in soups or on their own as a flavorful side.

All the varieties can be found in supermarkets and health food stores, both in dried and canned form. I keep Great Northern and cannellini beans in my house for making hummus or topping salads. The creamy and earthy texture of cannellini beans makes a great bean salad, while Great Northern beans work well when paired with other grains and flavors.

White Bean Flour

Out of all the flours, white bean flour is my go-to flour for creating creamy soups and thick sauces. The mild flavor of the bean blends well with other flavors. White bean flour also makes an excellent addition to gluten-free flour mixes, because it melds with the other flour flavors and helps create a moist texture in baked goods.

Any of the varieties of white beans can be ground into flour, but I've found that Great Northern are the easiest and have the most pleasing flavor. Cannellini beans are slightly too large for some electric grain mills to handle, and navy beans have a slightly different flavor. I shy away from grinding marrow beans because of their stronger flavor profile. However, the flavor could be a nice addition to soups or stews, depending on the other ingredients. Experiment with all three to find the flour that best fits your needs.

Depending on size, white beans are best ground in a grain mill or high-speed blender. If I am grinding cannellini beans, I often use my coffee mill and sift the flour through a sieve, because the beans are slightly too large for the grain mill chute.

Weights and Measurements

1 cup white beans = 180 g

1 cup white bean flour = 120 g

1 cup (180 g) white beans = 1½ cups (180 g) white bean flour

CHEESY BROCCOLI SOUP

If you were to peruse the archives of my website, one thing
would become abundantly clear: I love cheese, and I especially love cheesy vegetables.
Broccoli Cheddar soup was one of the first things I learned to make for myself
after falling in love with it at a restaurant. While I have a couple of different
recipes, I love the extra creaminess the white bean flour creates
with the broth in this version.

YIELD: 2 SERVINGS

1 tablespoon (15 ml) olive oil

1 small (80 g) onion, minced

1 shallot, minced

2½ cups (588 ml) vegetable broth

2 tablespoons (15 g) white bean flour

2 cups (180 to 200 g) ½-inch (1.2-cm)
   broccoli florets

1½ cups (135 g) shredded sharp Cheddar,
   plus more for serving (optional)

In a saucepan over medium-low heat, heat the olive oil. Add the onion
and shallot and cook until translucent, 6 to 8 minutes.

Add the vegetable broth and white bean flour, whisking until combined
and no flour lumps are left. Bring to a boil, and then reduce the heat to a
simmer. Add the broccoli to the saucepan and cook for 10 to 15 minutes,
until tender.

Remove the saucepan from heat and, using an immersion or regular
blender, purée the soup so that the broccoli is in small pieces. Return
the soup to the saucepan, add the cheese, and stir until melted, heating
again only if needed to melt the cheese.

Ladle the soup into bowls and serve with more cheese, if desired.

WHITE BEAN "HUMMUS"

There are plenty of recipes in cookbooks and on the Internet
for hummus, some with the end result of beautifully smooth hummus—which is
accomplished by removing the skin of the chickpeas. This is my way of working around
having to peel beans: By using bean flour, the hummus comes out smooth. Using
white bean flour instead of chickpea flour helps mellow the bean flour taste.

YIELD: 3 CUPS (668 G)

½ cup (60 g) white bean flour
2 cups (475 ml) water
1 clove garlic, minced
¼ cup (40 g) tahini
3 tablespoons (45 ml) fresh lemon juice
3 tablespoons (45 ml) olive oil
½ teaspoon sea salt

In a medium saucepan, whisk together the white bean flour and water.
Turn the heat to medium-low and continue to stir as the mixture cooks
and thickens, 6 to 8 minutes. The flour taste should be cooked out and
there should be no lumps.

Transfer the mixture to a food processor or blender. Add the garlic, tahini,
lemon juice, olive oil, and salt and pulse to combine and loosen the bean
mixture. If the hummus is too thick, add a bit more water, lemon juice, or
olive oil to suit your taste.

This hummus doesn't quite have the same texture as traditional hummus, but
I find it tastes just as good, and it's easier to make smooth.

GORGONZOLA
MAC AND CHEESE

We are not big milk drinkers in my house, and often I'll reach
for milk to use in a recipe only to realize it's expired. Using white bean flour to help
thicken a sauce is the perfect way to create a makeshift mac and cheese dinner.

YIELD: 4 SERVINGS

8 ounces (225 g) bow tie or penne pasta

¼ cup (30 g) white bean flour

½ teaspoon sea salt

2 cups (475 ml) water

3 ounces (85 g) Gorgonzola

2 teaspoons fresh lemon juice

½ cup (16 g) grated Parmesan

½ cup (60 g) bread crumbs (optional)

Preheat the oven to 375°F (190°C, or gas mark 5).

Bring a medium pot of salted water to a boil. Add the pasta and cook
until just shy of al dente. Drain and set aside.

Return the pot to the stove top and add the bean flour, salt, and water,
whisking until the mixture is smooth and there are no lumps of flour.
Bring to a boil, then reduce the heat to a simmer and let cook until the
mixture has thickened and flour taste has cooked out, 6 to 8 minutes.

Remove from the heat and add the Gorgonzola and lemon juice. Let sit
for 1 to 2 minutes to melt the cheese, and then stir until the cheese is
melted into the sauce.

Combine the cooked pasta with the cheese sauce in a 2-quart (1.9-L)
baking dish. Top with the Parmesan and bread crumbs, if using. Bake
for 25 to 30 minutes, until the cheese sauce is bubbling and the top is
lightly browned.

If Gorgonzola isn't your favorite, try using a sharp Cheddar instead.

CHAPTER 5

Nourishing Nuts and Seeds

Nut and seed flour or meal is relatively new in its use as a replacement for traditional grain flour. Nut meals and flours have garnered new attention thanks to the Paleo diet, which excludes grains and legumes. Since I'm not on this diet, I tend to mix nut and seed meals with gluten-free flours. While I love nut meals, I find that the quantity it sometimes takes to make baked goods can be a bit much, both in richness and in flavor. My main exception to this is the Lemon Pistachio Cake on page 197—it's a perfect celebration cake, as the sweet, nutty flavor of the pistachios pairs beautifully with the brightness of the lemon.

Although most seeds are fine stored at room temperature, nuts should be refrigerated to keep them fresh. I've had numerous experiences of buying nuts from bulk bins only to find, once home, that the nuts were stale, almost rancid. I try to buy nuts from stores that keep the nuts in refrigerated sections. If that is not an option, I politely ask to sample the nut to ensure freshness.

Nuts and certain seeds also make for wonderful butters, which is part of the reason I keep so many different varieties on hand. However, as exciting as nut butters can be, they can become a giant pain when grinding nut flours. Nuts and seeds have a high fat content that creates the creaminess of the butters. The fat also releases when you grind nuts into flours, and it is easy to go from nut meal to nut butter very quickly! Grind small batches of nuts in the food processor or coffee grinder, pulsing only a few times.

If you buy nut and seed flours from a store, you may notice that some ground nuts and seeds are marked "meal" while others are marked "flour." For grinding flour at home, nut and seed *flour* is what results after sifting the *meal*, or the first grind. In some recipes, meal and flour can be used interchangeably, while in others, it can result in an unsatisfactory dish. I provide specific information for each recipe.

Finally, fresh nuts make wonderful nut milk. There are endless amounts of nut milk recipes online that are as simple as soaking, blending, and straining nuts. One of the benefits of making nut milk is the leftover pulp, which can be dried in a 200°F (100°C, or very low) oven for 2 to 3 hours, then grinded in a food processor or blender to make nut meal.

FLAXSEEDS

In my earlier years of eating junk food, any mention of flaxseed or flaxseed oil conjured mental images of overly healthy food, of which I wanted no part. Some time later, I started experimenting with alternative baking, usually vegan, and I learned that one of the best substitutes for an egg is flaxseeds mixed with water. I discovered that while flaxseeds have a good amount of nutrients, the tiny seeds have a wonderful and distinct sweet, earthy flavor. From that moment, flaxseeds started making their way into my granolas, muffins, and even breads.

Flaxseeds, also known as linseeds, are cultivated around the world for various reasons. Before cotton, there was flax, which was spun into linen. Flaxseeds are also pressed into oil for human consumption, for uses in medicine and cooking, and for nonhuman consumption, such as for use in paints, soaps, and furniture oils.

Flaxseeds for consumption are prized for their high levels of omega-3 fatty acids and fiber, both of which help keep our bodies functioning properly. However, because whole flaxseeds pass through the body without absorption of their nutrients, ground flaxseeds are a great way to reap all their nutritional benefits.

Flaxseeds are primarily available in two colors: golden and brown, but be on the lookout for other colors, such as white or black. Such seeds were harvested too early or too late, so don't buy them. The nutritional makeup of the golden and brown seeds is similar. Brown and golden flaxseeds can be purchased prepackaged or in the bulk bins at health food stores and occasionally prepackaged in grocery stores.

Flaxseed Meal

Grinding flaxseeds into meal is best done in a high-speed blender or coffee grinder. The seeds are too oily for an electric grain mill and too fine for a food processor. As for color, I'm not terribly picky and buy whatever is sold at my local store. Both varieties of flax grind well and have similar nutritional benefits. Golden flaxseeds, however, have a slightly nuttier, sweeter flavor, while the brown flaxseeds are a bit milder.

Though many stores sell flaxseed meal, this is one thing I always grind myself. Ground flaxseeds have an extremely short shelf life, a mere week or two if left in a cupboard. I grind whatever I need at the time, and if I happen to have extra meal left over, I always refrigerate it and smell the meal before using. If the meal has a foul, slightly fishy odor, pitch it.

When using flaxseed meal, be aware that sometimes the texture may be slightly more gelatinous when used in porridge and other nonbaked goods. Flaxseeds and meal, when combined with water, begin to expand and become gummy. While I do not think it's an unpleasant texture, it may feel a bit different when you're new to using flax.

Weights and Measurements

1 cup flaxseeds = 150 g

1 cup flaxseed meal = 100 g

1 cup (150 g) flaxseeds = 1½ cups (150 g) flaxseed meal

FLAXSEED MEAL—CRUSTED
AVOCADO SPRING ROLLS

I am a spring roll fanatic, especially during the summer months,
when produce is flowing. Mostly my spring rolls are chock-full of raw vegetables,
but sometimes I mix it up by grilling or roasting. Crusting the avocado
in the flaxseed meal gives the spring roll a bit of extra flavor
and makes for a fun addition.

YIELD: 8 SPRING ROLLS

2 barely ripe avocados, halved and pitted

1 cup (100 g) flaxseed meal

1 tablespoon (15 ml) sesame oil

5 cups (350 g) shredded red cabbage

1 large red bell pepper, cut into thin strips

¼ cup (4 g) fresh cilantro

3 tablespoons (45 ml) fresh lime juice

1 tablespoon (20 g) honey

8 rice paper spring roll wrappers

Soy sauce, for serving

Carefully run a spoon between the skin and the flesh of the avocados
and scoop each half out onto a cutting board. Slice each half into ¼-inch
(6-mm) strips. Pat the avocado strips into the flaxseed meal so that the
avocado is coated.

In a large skillet over medium-low heat, heat the sesame oil. Place the
avocados in the skillet and cook on each side until golden and crisp,
1 to 2 minutes per side.

While the avocados are cooking, combine the cabbage, red pepper, and
cilantro in a large bowl. In a small bowl, whisk together the lime juice
and honey. Pour the mixture over the vegetables and toss until the
cabbage is coated.

Set up a rolling station with the cooked avocados, cabbage mixture,
spring roll wrappers, a dish of hot water large enough to hold the spring
roll wrappers, and a cutting board.

Soak each wrapper for 10 to 15 seconds. The wrapper should be pliable
but not so soft that it's hard to remove from the water. Place the wrapper
on the cutting board with a point toward the bottom, closest to you (if
using square wrappers). Add ⅓ cup (69 g) cabbage mixture to the lower
portion of the wrapper and 2 or 3 slices of avocado. Roll the wrapper up,
tucking and folding in the sides as you go. Set aside. Continue with the
remaining ingredients.

Cut the spring rolls in half and serve with a small bowl of soy sauce.

MINI SPINACH QUICHES WITH FLAX CRUST

In recent years, quiche has made an appearance at every
family holiday. It started out as a way to serve a substantial vegetarian
main dish but has turned into a tradition. If I'm having people over for breakfast,
quiche is my first go-to recipe. Using flaxseed meal instead of traditional
piecrust is a fun way to add a bit of crunch to the quiches.

YIELD: 6 MINI QUICHES

For the crust:

1¼ cups (125 g) flaxseed meal

1 large egg white

1 tablespoon (15 ml) olive oil

1 teaspoon honey

For the filling:

4 large eggs

1½ cups (360 ml) 2% or whole milk

½ teaspoon garlic powder

¼ teaspoon sea salt

¼ teaspoon freshly ground black pepper

1½ cups (40 to 50 g) loosely packed
  spinach, chopped

¾ cup (68 g) shredded fontina

Preheat the oven to 375°F (190°C, or gas mark 5). Line a jumbo 6-cup muffin pan with paper liners.

To make the crust: In a medium bowl, combine the flaxseed meal, egg white, olive oil, and honey. Divide the mixture among the muffin cups and press into the bottom and sides. Bake for 15 to 18 minutes, until crisp and golden. Let cool slightly.

To make the filling: Whisk together the eggs, milk, garlic powder, salt, and pepper. Divide the spinach among the cooled crusts. Sprinkle 2 tablespoons (11 g) of the cheese over the spinach. Finally, pour the egg mixture over the spinach and cheese.

Return to the oven and bake for 20 to 25 minutes, until the egg mixture is set and puffed. Serve immediately.

Because of the crisp nature of this crust, it's easier to manage in the smaller servings. This quiche could be made in a 9-inch (23-cm) pie pan, but solid pieces that hold together may be hard to cut.

FLAX PORRIDGE WITH PEACHES

During the cooler months, my mornings always consist of a
nice brisk walk, a steaming cup of coffee, and a bowl of hearty porridge.
I like to mix my porridge up by using different grains and occasionally seeds.
The consistency of this porridge is rather fun, since flaxseeds gel with liquid,
but the end result is still a hearty and nutritious breakfast.

YIELD: 2 SERVINGS

¼ cup (25 g) flaxseed meal

1 to 1½ cups (235 to 355 ml)
   2% or whole milk

2 ripe peaches, sliced

2 tablespoons (40 g) honey

1 tablespoon (14 g) butter

In a medium saucepan, whisk together the flaxseed meal and 1 cup
(235 ml) of the milk. Bring to a boil, then reduce the heat to a simmer,
whisking frequently. Cook the porridge until thick, 4 to 5 minutes, and
add more of the milk to thin the porridge, if desired.

Meanwhile, in a skillet over medium heat, cook the peaches with the
honey and butter for 3 to 4 minutes, until tender. Serve the porridge
topped with the peaches.

Occasionally, instead of making a porridge completely from flaxseed meal, I add
1 to 2 tablespoons (6 to 12.5 g) of the meal to other porridges, such as amaranth,
millet, and even oatmeal.

PUMPKIN SEEDS

Pumpkin seeds, especially roasted, are one of my favorite snacks. While carving pumpkins one year, a friend looked confused as to why I was meticulously pulling out the seeds and setting them aside. She questioned why it looked like I cared more about the seeds than I did about the actually carving—which, to a certain extent, was true. One of the best parts about carving pumpkins is roasting the seeds with a pinch of salt and smoked paprika. I always have a bag of pumpkin seeds around the house to toss onto salads, top soups, or grind into flour to make lovely treats such as the cupcakes on page 181. One of my favorite traditions in the autumn is to purchase a few pumpkins to stock up on pumpkin purée and pumpkin seeds for snacking.

The pumpkin seeds that are pulled fresh from pumpkins are covered with a hull. This version of the seed can be eaten, especially when roasted, but the majority of pumpkin seeds sold in stores are without the hull. The distinction between the two is apparent, as the hull is oval shaped and tan while the pumpkin seed is green and slightly skinnier. Pumpkin seeds are a good source of zinc, protein, and iron.

When purchasing pumpkin seeds, there are many options. I usually keep raw, shelled pumpkin seeds on hand for grinding into flour, tossing into granola, or roasting for a snack. Look for pumpkin seeds in the bulk bins and prepackaged sections of health food stores and some supermarkets. Also, do not be confused by the label "pepita," as this is simply their Spanish culinary term.

Pumpkin Seed Meal

Pumpkin seed meal is a fun addition to meals and a useful replacement for nut meals. However, I love using pumpkin seed meal for binding ingredients together, as in the lentil "meat" balls on page 179.

The meal is best ground in a coffee grinder or high-speed blender. While you can sift pumpkin seed meal into flour, I find the meal is finely ground enough that I usually do not need to sift it. After grinding, the meal has a somewhat unpleasant odor, but do not let that deter you from using it. The meal, when cooked, is mild. Pumpkin seeds and meal do go rancid quickly. Because of this, I often buy small quantities of pumpkin seeds and only grind what I need.

Weights and Measurements

1 cup pumpkin seeds = 120 g

1 cup pumpkin seed meal = 100 g

1 cup (120 g) pumpkin seeds = 1 cup plus 3 tablespoons (120 g) pumpkin seed meal

LENTIL "MEAT" BALLS

I was never a big meatball eater. If I had the option of ordering
spaghetti with or without meatballs, I would always opt for no meatballs. Yet
when I first tried my hand at lentil "meat" balls, I fell in love. They make for a filling
vegetarian main dish that works whether smothered in barbecue sauce or
buffalo sauce or paired with the spaghetti recipe on page 74.

YIELD: 8 LARGE "MEAT"BALLS

1 tablespoon (15 ml) olive oil

⅓ cup (60 g) minced onion

1 small clove garlic, minced

1 cup (160 g) cooked brown lentils, cooled

½ cup (50 g) pumpkin seed meal

¼ cup (25 g) pumpkin purée

1 large egg yolk

2 teaspoons minced fresh parsley

1 teaspoon minced fresh sage

½ teaspoon sea salt

½ teaspoon freshly ground black pepper

Preheat the oven to 375°F (190°C, or gas mark 5). Line a baking sheet with parchment paper.

In a medium skillet over medium heat, heat the olive oil. Add the onion and garlic, cooking until the onion is translucent, 5 to 6 minutes. Let cool slightly.

In a large bowl, combine the lentils, pumpkin seed meal, onion mixture, pumpkin purée, egg yolk, parsley, sage, salt, and pepper. Stir until well combined and the mixture can be formed into balls that hold together.

Scoop out roughly ¼ cup of the lentil mixture and shape it into a ball. Place on the prepared baking sheet. Continue with the remaining mixture to make 8 balls.

Bake the "meat"balls, without turning them, until the outside is crisp and slightly golden, 40 to 45 minutes. They can be made ahead of time and stored in the freezer until ready to bake; just let them thaw before baking them.

The mixture can be used to make smaller "meat"balls, perfect for an appetizer. Simply scoop out 2-tablespoon (21 g) portions of the mixture to make 16 meat-balls, and adjust the cooking time to about 30 minutes.

CHOCOLATE CUPCAKES
WITH GANACHE

I like to call these my "surprise" cupcakes because people are
always surprised that one of its main ingredients is pumpkin seeds. These
bake up moist, and the pumpkin seeds add just a hint of earthiness. For extra flavor,
try adding 1 to 2 tablespoons (9 to 18 g) of espresso powder to the batter.

YIELD: 12 CUPCAKES

For the cupcakes:

2 cups (200 g) pumpkin seed meal

½ cup (40 g) unsweetened cocoa powder

½ teaspoon baking soda

¼ teaspoon sea salt

4 large eggs

¼ cup plus 2 tablespoons (125 g) honey

¼ cup (60 ml) walnut oil

For the ganache:

¼ cup (60 ml) heavy cream

1 cup (224 g) semisweet chocolate chips

To make the cupcakes: Preheat the oven to 375°F (190°C, or gas mark 5). Line a standard 12-cup cupcake pan with liners.

In a large bowl, stir together the pumpkin seed meal, cocoa powder, baking soda, and salt. In a separate bowl, whisk together the eggs, honey, and walnut oil. Pour the wet ingredients into the dry ingredients and stir until the batter is smooth. Divide the batter among the cupcake wells, filling each about two-thirds full.

Bake for 16 to 18 minutes, until the cupcakes spring back when pressed. Transfer to a wire rack and let cool.

To make the ganache: In the top of a double boiler, combine the heavy cream and chocolate chips. Heat and stir until the chocolate melts and the ganache is smooth.

Once the cupcakes have cooled, spread the ganache on top and serve.

STUFFED MUSHROOMS WITH PUMPKIN MEAL TOPPING

Mushrooms are one ingredient I never fell head over heels
in love with. In the past few years, I've gone from "avoiding them at all cost"
to "if they happen to be there, I might eat them." My husband, on the other hand,
loves them, and I try to keep a few mushrooms in the house for salads, stir-fries,
and grilling. These stuffed mushrooms are a special treat for him.

YIELD: 12 MUSHROOMS

2 ounces (55 g) goat cheese
2 ounces (55 g) cream cheese, softened
½ teaspoon garlic powder
½ teaspoon freshly ground black pepper
¼ teaspoon sea salt
12 large button mushrooms, stems removed
¼ cup (25 g) pumpkin seed meal
¼ cup (8 g) grated Parmesan

Preheat the oven to 375°F (190°C, or gas mark 5).

In a bowl with a wooden spoon, mix together the goat cheese and cream cheese. Add the garlic powder, pepper, and salt and stir to combine. Using a butter knife or spoon, fill the mushrooms with 1 to 2 tablespoons (9 g) of the cheese mixture, mounding it if necessary.

Combine the pumpkin seed meal and Parmesan in a bowl. Dip the tops of the mushrooms in the pumpkin meal mixture, coating the cheese mixture, and place in a baking pan. Bake the mushrooms for 20 to 25 minutes, until the mushrooms are soft and the topping is slightly golden.

SUNFLOWER SEEDS

When I first moved to California, I was in constant amazement at all the fresh produce that grows in the Sacramento Valley. It seemed, quite possibly, that I could get whatever I needed from this area. I love driving through growing country, and one of my favorite sights is an endless field of sunflowers. Sunflower seeds are a hidden gem. The flavor is sweet on its own, makes for wonderful additions to meals, and is lovely as sunflower seed butter (a great replacement for those who have nut allergies). I keep sunflower seeds in my house for tossing on salads, for adding crunch to veggie burgers, and for raw crusts such as the one on page 187.

Sunflower seeds have two parts: the kernel and the hull. The teardrop-shaped hull is the hard, outer shell that surrounds the tan-colored kernel. The hull has distinct features that helps determine the seed's use. If the hull is black, the sunflower seed is called black oil and is pressed for oil or used for bird feed. If the outer hull is black with a few white stripes, it is referred to as the confectionary seed, used for human consumption.

Sunflower seeds are sold with and without the hull. I buy raw sunflower kernels, not those with the hull, to grind into meal; shelling becomes tedious after just a few! Look for sunflower seeds prepackaged or in the bulk-bin sections of health food stores and supermarkets.

Sunflower Seed Meal

Similar to many nuts, sunflower seeds have a high oil content, which is great for making nut butters but can make meal grinding a bit tricky. I prefer to pulse sunflower seeds in a coffee grinder to get meal without turning them into butter. Food processors and blenders work as well, but I find it is a delicate dance to get a fine meal before turning it into butter. Similar to flaxseed meal and pumpkin seed meal, sunflower seed meal grinds fine enough to not require sifting into flour.

I enjoy using sunflower seed meal, as the slightly sweet taste of the seed stays in the meal. Everything I have made from sunflower seed meal has a pleasant, sunny taste that always has a hint of the slightly nutty sunflower seeds. The sweetness of the meal makes it a perfect nut-free substitution for almond meal.

If you use sunflower seed meal with baking soda, use caution. A chemical reaction between the two may cause your baked goods to have a green hue. This reaction is perfectly safe and doesn't affect the taste, but the sight may be a bit unappealing.

Weights and Measurements

1 cup sunflower seeds = 100 g

1 cup sunflower seed meal = 100 g

1 cup (100 g) sunflower seeds = 1 cup plus 2 tablespoons (100 g) sunflower seed meal

SUNFLOWER SEED CRACKERS

I have many favorite snack items, including cheese and crackers.
However, crackers from the store often contain extra ingredients for added shelf life
and can occasionally contain unnecessary sugar. These crackers have minimal ingredients,
are easy to throw together, and make for the perfect snack base.

YIELD: 24 TO 36 CRACKERS, DEPENDING ON DESIRED SIZE

1 cup (100 g) sunflower seed meal
½ cup (50 g) hazelnut meal
½ cup (50 g) pecan meal
1 large egg
1 teaspoon sea salt

Preheat the oven to 375°F (190°C, or gas mark 5). Line 2 baking sheets with parchment paper.

In a bowl, stir together the meals, egg, and sea salt until well combined. Divide the mixture in half and set one half on each baking sheet. Cover with more parchment paper or waxed paper and roll each piece as thin as you can make it, almost paper-thin. Remove the top piece of parchment, then cut the crackers into 1 × 2-inch (2.5 × 5-cm) rectangles using a knife or pizza cutter. Do not separate the pieces. Bake for about 8 minutes.

Remove from oven and, using a knife or pizza cutter, separate the crackers. Return the crackers to the oven and bake for another 3 to 5 minutes, until crisp. Let cool before serving.

BANANA CREAM PIE WITH RAW SUNFLOWER CRUST

During the holidays, my father's favorite pie was chocolate cream,
and because of this, we would have it every year. When I started to work at a bakery,
I learned the joys of branching away from chocolate into banana and coconut, and
I fell in love with banana cream pie. This is my twist, using a raw seed and nut crust, and
honey in place of the large amount of sugar usually called for in cream pies.

YIELD: 8 SERVINGS

For the crust:

6 to 8 (80 g) Medjool dates, pitted
1 cup (100 g) sunflower seed meal
½ cup (50 g) almond meal
½ cup (50 g) hazelnut meal
¼ teaspoon ground cinnamon
¼ teaspoon sea salt
1 to 2 tablespoons (20 to 40 g) honey

For the filling:

2 cups (475 ml) whole milk
4 large egg yolks
⅓ cup (112 g) honey
¼ cup (32 g) cornstarch
¼ teaspoon sea salt
1 teaspoon vanilla extract
2 large ripe bananas, sliced

For the topping:

1½ cups (360 ml) heavy cream
2 teaspoons honey

To make the crust: Lightly grease a 9-inch (23-cm) pie pan. In a small bowl, cover the dates with warm water and let sit until soft, 20 to 30 minutes. Drain them and combine with meals, cinnamon, and salt in a food processor. Pulse until the dates are in small pieces. Spoon in the honey and pulse until the mixture starts to clump together, adding a little more honey if needed. Press the crust into the prepared pie pan and place in the refrigerator until ready to fill.

To make the filling: In a medium saucepan, heat the milk over medium-low heat until the milk is warm and starts to bubble around the edges. In a medium bowl, whisk together the egg yolks, honey, cornstarch, and salt until smooth. Pour ¼ cup (60 ml) of the warm milk into the egg mixture to warm the eggs. Then, while continuing to whisk the egg mixture, slowly, in a steady stream, pour in the remaining milk. Once combined, pour the filling back into the saucepan and return to medium-low heat.

Heat the mixture, continuing to whisk, until it reaches a pudding consistency, 3 to 5 minutes. As the filling heats, it may clump; just continue to whisk. Once thickened, remove from the heat and whisk in the vanilla.

Cover the bottom of the chilled crust with the sliced bananas, layering as needed. Pour the cooked filling over the bananas and place the pie in the refrigerator to cool completely, for 1 to 2 hours.

To make the topping: Whip the heavy cream, either by hand or with a mixer, until it forms stiff peaks; stir in the honey. Smooth the whipped topping over the pie and serve. Store any left over in the refrigerator.

SUNFLOWER JAM COOKIES

I take after my father in a lot of ways. I followed in his footsteps
playing hockey and learning photography; we both have slightly stubborn personalities
and a love of being outdoors; and we both have a weakness for my grandmother's
thumbprint cookies. This recipe is a riff on those thumbprint cookies, using sunflower
seed meal and jam in place of the icing she would normally use.

YIELD: 12 COOKIES

1 cup (100 g) sunflower seed meal
¼ cup (34 g) sorghum flour
¼ cup (30 g) buckwheat flour
2 tablespoons (16 g) arrowroot
2 tablespoons (16 g) cornstarch
2 tablespoons (28 g) butter, melted
1½ tablespoons (30 g) honey
½ teaspoon vanilla extract
2 to 3 tablespoons (40 to 60 g)
   strawberry, blueberry, or raspberry jam

Preheat the oven to 350°F (180°C, or gas mark 4). Line a baking sheet with parchment paper.

In a bowl, stir together the sunflower seed meal, flours, arrowroot, and cornstarch. Add the butter, honey, and vanilla. Stir until a dough forms.

Using a small scoop or spoon, divide the dough into 12 mounds. Roll each mound into a ball and place on the baking sheet. Using your thumb, press down in the center of each cookie, creating an indent. Spoon ½ to 1 teaspoon jam in the center.

Bake for 12 to 13 minutes, until the cookies are firm and slightly browned. Let cool before serving. Store in an airtight container at room temperature for 2 to 3 days, or freeze for extended storage.

ALMONDS

Next to pistachios, almonds are my standby snack nut. I turn to almonds when I'm looking for a light, buttery taste that always pairs well with dried fruit, olives, cheese, and, of course, wine. In addition, almonds are my hiking snack for a boost of quick energy on the trail. These nuts are all-around great, and I always have a stash on hand.

Almonds, though labeled a nut in the culinary world because of their similar use to true nuts such as walnuts and hazelnuts, are actually the fruit of a deciduous tree that is gorgeous when in bloom. The almond grows with three distinct layers: an outer leathery green hull, a hard shell, and the seed, which is what we know as the almond. These nuts originated in the Mediterranean and Middle East but are now grown around the world, including my current home state of California.

There are two main varieties of almonds: sweet almonds, which are the ones we consume, and bitter almonds, which are made into almond oil. Almonds can be purchased in bulk bins or prepackaged at most supermarkets and health food stores. Almonds are high in fiber and protein as well as a good source of healthy unsaturated fats.

Almond Meal and Flour

The variety of almonds at the store can be a bit overwhelming. There are plain (raw), roasted, plain salted, roasted salted, and more. Almonds can also be blanched and slivered or sliced, with or without skins. My preference is to stick with the unsalted plain or roasted almonds, rather than blanched, since I keep them around for a snack. Also, they both make great almond meal and flour with gorgeous brown specks.

Almonds have a little less fat than other nuts, which makes them perfect for grinding into delicate flour (although it can also be further ground into a delicious butter). Pulse almonds in a food processor, shake through a sieve, and return the almond meal to the food processor. Repeat this process until the majority of meal has been pulsed and sifted into flour. Almond flour can be used alone or substituted in recipes for part of the flour, and the leftover meal is great in the granola on page 190.

Weights and Measurements

1 cup almonds = 140 g

1 cup almond meal = 120 g

1 cup (140 g) almonds = 1¼ cups plus 2 tablespoons (140 g) almond meal

Almond Meal Honey Granola

A perfect breakfast for me is whole-milk Greek yogurt,
fresh fruit, and a sprinkle of granola. I love using almond meal for granola
because it adds a little extra protein punch to start your day!

YIELD: 4 CUPS (479 G)

2 cups (240 g) almond meal

1 cup (130 g) dried apricots, chopped

1 teaspoon ground cardamom

¼ teaspoon sea salt

3 tablespoons (45 ml) walnut oil or melted butter

3 tablespoons (60 g) honey

Preheat the oven to 300°F (150°C, or gas mark 2).

In a medium bowl, combine the almond meal, apricots, cardamom, and salt.

In a separate bowl, whisk together the oil and honey. Pour over the almond meal mixture and stir until the almond mixture is coated. Spread the granola in a thin layer on a baking sheet. Bake for 25 to 30 minutes, stirring once around the 15-minute mark. The almond meal should be golden.

Let cool completely; the granola will crisp as it cools. Store in an airtight container at room temperature for up to 1 week.

The apricots can easily be changed out for another type of dried fruit, such as raisins, cranberries, or even pineapple.

DARK CHOCOLATE–DIPPED ALMOND DROP COOKIES

These cookies are what I call a happy accident. I whipped them up
one day when I was craving a cookie but had no butter and no gluten flour.
The result was a light and airy cookie that can satisfy any sweet tooth. Be sure to
use almond flour, as the cookies will not turn out well with almond meal.

YIELD: 12 COOKIES

1 cup (120 g) almond flour

1 teaspoon ground cinnamon

1 teaspoon baking soda

¼ teaspoon sea salt

2 egg whites

2 tablespoons (30 ml) walnut oil or
melted butter

4 teaspoons (60 ml) maple syrup

½ teaspoon vanilla extract

⅔ cup (117 g) chunked dark chocolate

Almond meal, for garnish

Preheat the oven to 350°F (180°C, or gas mark 4).

In a medium bowl, whisk together the almond flour, cinnamon, baking
soda, and salt. Add the egg whites, oil, maple syrup, and vanilla, stirring
until the dough is thoroughly mixed.

Using a spoon or cookie scoop, drop cookies onto a baking sheet in
2-tablespoon (23 g) portions and press into a circle. Bake for 18 minutes,
until puffed up and slightly firm to the touch. Let cool completely.

In the top of a double boiler over medium-low heat, melt the dark
chocolate. Dip half of each cookie into the chocolate and place on a
drying rack or parchment paper. Sprinkle almond meal over the chocolate
and let sit until the chocolate hardens.

HONEY ALMOND CREPES WITH ROASTED PEACHES AND WHIPPED CREAM

During the peak of summer produce season, I try to
find ways to incorporate fresh produce into everything. The almond
crepes and roasted peaches play wonderfully together, and the lightness
of the crepes is the perfect finish to those warm summer nights.

YIELD: 6 TO 8 CREPES

¾ cup (90 g) almond flour

2 tablespoons (16 g) arrowroot

⅛ teaspoon sea salt

1 large egg

¼ cup plus 2 tablespoons (90 ml)
   2% or whole milk

2 tablespoons (40 g) honey, divided

1 tablespoon (14 g) butter, melted

Walnut or coconut oil, for the pan

3 or 4 ripe peaches, sliced

½ cup heavy cream

Preheat the oven to 375°F (190°C, or gas mark 5).

In a medium bowl, whisk together the almond flour, arrowroot, salt, egg, milk, 1 tablespoon (20 g) of the honey, and the butter until smooth. Heat an 8-inch (20-cm) skillet over medium-low heat and lightly grease with oil. Pour a scant ¼ cup (48 g) of batter in the skillet. Working quickly, tilt the skillet in a circle so that the batter covers the entire bottom and cook for about 30 seconds. Flip and cook for another 15 seconds. Layer finished crepes, slightly overlapping, on a plate.

Place the peaches in a single layer in a roasting pan. Bake for 15 to 25 minutes, until tender (riper peaches will take less time). While the peaches roast, whip the heavy cream until soft peaks form.

Scoop roughly ⅓ cup of the peaches on one half of the crepe, fold over, and top with another spoonful of the peaches. Serve with a drizzle of the remaining honey and the whipped cream.

Crepes freeze well, and I often make up a batch and use only 1 or 2 at a time. To freeze, simply separate the crepes with parchment paper and seal in a freezer-safe bag.

PISTACHIOS

I only started eating pistachios recently, when I forgot my snacks on a hike and shared my mother's. After my first few bites of the pistachios, I wondered how I had gone so long without eating anything containing pistachios (minus the occasional treat of baklava from the local bakery). From then on, pistachios became a regular snack in my house, usually shelled with a bit of salt. I also am quite fond of pairing pistachios with chocolate and occasionally lemon. Pistachios have a unique, slightly sweet flavor with a lightness that is rivaled only by almonds.

What we consider the pistachio nut is actually the kernel from the seed of a small flowering fruit tree. The kernels are grown in the safety of a husk and hard, tan shell, which can be opened easily enough by hand. Pistachio nuts are lovely: green with a thin purple skin. The pistachio is part of the cashew family, and both are related to poison ivy and sumac. Pistachios originated in the Middle East and Central Asia but are now grown around the world. They are highly marketed as a snack nut, sold both in and out of the shell. Pistachios are an excellent source of protein and high in healthy fats.

Pistachios in the shell can be purchased at supermarkets and health food stores, both prepackaged and in bulk bins. Pistachios are more often than not sold roasted and salted but can be found raw. I've found that hunting down shelled raw pistachios isn't always easy, but they can usually be found in the bulk-bin sections of health food stores or from online companies.

Pistachio Meal and Flour

When purchasing pistachios to make meal and flour, buy the unshelled, as shelling takes time. If I know that I'm going to make a recipe that requires pistachio meal, I will purchase unsalted raw pistachio nuts without the shell. Always look for the unsalted, as salted pistachios have a hefty amount of salt that can ruin a recipe.

This flour is one of the only nut flours I use on its own in baked goods. I use a food processor to grind the pistachios in small batches to ensure I do not end up with pistachio butter. When sifted, pistachio meal creates a silky green flour; just be aware that this often results in green baked goods.

Pistachio meal is one of my favorite additions to recipes. I enjoy subbing pistachio meal as part of the total amount of flour in muffin recipes and adding pistachio meal and milk to vanilla ice cream. I also sometimes sprinkle pistachio meal instead of granola over yogurt.

Weights and Measurements

1 cup pistachios = 100 g

1 cup pistachio meal = 100 g

1 cup (100 g) pistachios = 1 cup (100 g) pistachio meal

PISTACHIO-CRUSTED BUTTERNUT SQUASH SPINACH SALAD

I am a bit backward in my salad eating. Although most people stop eating salads once the weather cools, I increase my consumption, mainly because salad greens wonderful in the fall. This pistachio salad is a great way to have a warm salad that pairs the flavors of fall with fresh spinach.

YIELD: 2 SERVINGS

For the squash:

½ pound (225 g) butternut squash

½ clove garlic, minced

1 teaspoon fresh oregano

¼ teaspoon sea salt

¼ teaspoon freshly ground black pepper

1 tablespoon (15 ml) olive oil

¼ cup (25 g) pistachio meal

For the dressing:

2 ounces (55 g) blue cheese, crumbled

¼ cup (56 g) whole-milk plain Greek yogurt

2 tablespoons (30 ml) buttermilk

1 tablespoon (15 ml) apple cider vinegar

¼ teaspoon sea salt

¼ teaspoon freshly ground black pepper

2 to 3 handfuls spinach

To make the squash: Preheat the oven to 375°F (190°C, or gas mark 5). Lightly grease a baking sheet.

Peel the butternut squash and cut in half lengthwise. Remove the seeds and cut the squash into ¼-inch (6-mm) slices.

Place the slices in a medium bowl. In a small bowl, combine the garlic with the oregano, salt, and pepper. Sprinkle over the squash. Drizzle the olive oil over the squash and toss to coat. Add the pistachio meal to the squash, tossing a few more times until the squash is coated. Transfer in a single layer to the prepared baking sheet, patting any loose pistachio meal mixture onto butternut squash. Bake until the pistachio meal is slightly crisp and the squash is tender, 40 to 45 minutes.

To make the dressing: Combine the blue cheese, yogurt, buttermilk, vinegar, sea salt, and black pepper in a bowl or food processor. Stir until combined or pulse in a food processor to break down the cheese. Add more buttermilk if a thinner texture is desired.

Divide the spinach among 2 plates. Top each with half of the butternut squash and half of the dressing.

LEMON PISTACHIO CAKE WITH CREAM CHEESE FROSTING

Although I rarely make baked goods out of only nut flours,
this cake is my one special-occasion exception. I love the flavor of pistachios,
especially when paired with the lightness of lemon. This is a perfect
gluten-free celebration cake.

YIELD: ONE 8-INCH (20-CM) CAKE

For the cake:

2 cups (200 g) pistachio flour, plus some
  meal for garnish (optional)

½ cup (72 g) arrowroot

2 tablespoons (12 g) lemon zest

1 teaspoon baking soda

½ teaspoon sea salt

½ cup (120 ml) walnut oil

¼ cup (85 g) honey

3 large eggs

For the frosting:

6 ounces (170 g) cream cheese, softened

2 tablespoons (28 g) butter, softened

1 cup (120 g) confectioners' sugar

1 to 2 tablespoons (20 to 40 g) honey

2 to 3 tablespoons (30 to 45 ml)
  heavy cream

Preheat the oven to 350°F (180°C, or gas mark 4). Grease an 8-inch (20-cm) round pan with butter.

To make the cake: In a medium bowl, stir together the pistachio flour, arrowroot, lemon zest, baking soda, and salt. In a separate bowl, whisk together the walnut oil, honey, and eggs. Pour the wet ingredients into the dry ingredients and stir until combined.

Pour the batter into the prepared cake pan and bake for 20 to 22 minutes, until it is golden and has a dome. Let cool for 10 minutes.

Once cooled slightly, run a knife along the edges to loosen the cake. Flip the cake over onto a cake plate and finish cooling.

To make the frosting: Beat together the cream cheese and butter with a hand or stand mixer. Add the confectioners' sugar, 1 tablespoon (20 g) of the honey, and 2 tablespoons (30 ml) of the heavy cream. Continue to beat until the frosting is smooth. Taste and add more honey if desired. If the frosting is too thick to spread, beat in another tablespoon (15 ml) of heavy cream. Frost all the sides of the cake and sprinkle pistachio meal on top, if desired.

Sifting the pistachio meal into flour is key in this recipe. The cake made with pistachio meal is dense, while the cake made with pistachio flour is light and airy.

CHIPOTLE SWEET POTATO LATKES WITH CILANTRO DIP

In job interviews, whenever the question about naming
a weakness would be asked, I always had an answer ready: lack of patience.
This is why the majority of my cooking gets done in less than 30 minutes.
I love these latkes because most sweet potato recipes require cooking the
sweet potato before using it, while this recipe calls for just grating them. I've even
been known to leave the peel on and grate the entire potato.

YIELD: 6 TO 8 LATKES

For the latkes:

1 medium (8 ounces, or 225 g) sweet
potato, peeled

¼ cup (40 g) minced red onion

1 large egg

2 tablespoons (12 g) pistachio meal

¼ teaspoon baking powder

¼ teaspoon sea salt

¼ teaspoon ground chipotle chile powder

¼ teaspoon ground coriander

¼ teaspoon ground cumin

2 to 3 tablespoons (30 to 45 ml) coconut
or olive oil or Coconut oil, melted

For the dip:

¼ cup (56 g) mayonnaise

2 tablespoons (4 g) chopped fresh
cilantro, plus more for garnish

1 tablespoon (15 ml) fresh lime juice

1 teaspoon honey

¼ teaspoon sea salt

¼ teaspoon freshly ground black pepper

To make the latkes: Using the medium holes of a box grater, shred the
sweet potato. Using cheesecloth or a fine-mesh strainer, squeeze and
press the liquid from the sweet potato shreds. Combine the grated sweet
potato in a bowl with the red onion, egg, pistachio meal, baking powder,
salt, chipotle powder, coriander, and cumin. Stir until the batter is well
combined.

In a large skillet, heat the oil over medium low heat. Spoon 2 to 3
tablespoons (47 g) of the sweet potato mixture into the hot oil and press
down to form a ¼-inch (6-mm)-thick circle. Cook on each side for 4 to
5 minutes, until crisp and browning. Transfer to paper towels to drain.
Repeat with the remaining sweet potato mixture, adding more oil, if
necessary.

To make the dip: Stir together the mayo, cilantro, lime juice, honey, salt,
and pepper. Taste and adjust the flavors to your liking.

Serve the latkes with a small side of the dip and an extra sprinkle
of cilantro.

If you do not like mayo, trying using whole-milk Greek yogurt for the dip instead.

HAZELNUTS

When I started eating healthier and cooking more, I ended up becoming the cook for all family holidays and parties. Not that I minded. My love affair with nut meals started shortly after I picked up these responsibilities, when one spring day I decided to make a dark chocolate tart with a hazelnut crust. The crust—merely hazelnut meal, sugar, and heavy cream—when combined with a creamy chocolate filling became magical. Everything I had known about crusts felt shattered. Making crusts out of nut meal opened a new world to using nuts, and my exploration took off from that moment.

Hazelnuts, also known as cobnuts or filberts, are the tree nut of a deciduous tree called the hazel tree. There is documentation of differences between hazelnuts and filberts, including size and shape, but the names for the most part are used interchangeably, depending on regions. Similar to other tree nuts, hazelnuts have three parts: the husk, the shell, and the kernel, which we call the nut. As hazelnuts mature, the outer husk begins to crack open, making access to the nut easier. Like all other nuts, hazelnuts are a good source of protein and unsaturated fats.

The nut has a thin, dark skin that can sometimes be bitter, and many recipes recommend removing it. However, I am fine with the flavor and stopped removing the skin the first time I ruined a good towel trying to do so. (If you'd like to remove the skins before using or grinding, roast the hazelnuts for 10 minutes in a 375°F [190°C, or gas mark 5] oven. Let cool for 10 minutes, then rub in a towel that you're okay with staining.) Look for hazelnuts in the cooler section in health food stores, as hazelnuts go stale rather quickly. I like to ask for a sample before purchasing to ensure freshness.

Hazelnut Meal and Flour

While I am partial to paired hazelnut meal with chocolate, the meal and flour can have a wide array of uses in baked goods, granolas, and toppings. The goat cheese salad on page 201 quickly became a favorite. Hazelnut meal adds a subtle nuttiness to dishes and is also one of my favorite nut meals to toast. Place the meal in a skillet over medium-low heat, stirring for 3 to 4 minutes, until fragrant.

As I choose to grind my hazelnuts with the thin skin left intact, the result is a lovely speckled meal. Hazelnuts are best ground in the food processor in small batches. For hazelnut meal, simply pulse a few times. For hazelnut flour, pulse, sift, and return the hazelnut meal to the food processor. Hazelnut meal works great for coatings, like the truffles on page 202, and hazelnut flour is great for use in baked goods, especially muffins, pancakes, and brownies.

Weights and Measurements

1 cup hazelnuts = 120 g

1 cup hazelnut meal = 100 g

1 cup (120 g) hazelnuts = 1 cup plus 3 tablespoons (100 g) hazelnut meal

HAZELNUT-CRUSTED GOAT CHEESE SALAD

There was a short period of time when, if a restaurant where I was
eating featured a goat cheese salad, I would order it. I loved the warm, tender goat
cheese paired with the fresh crispness of the salad. I continued to order these salads until
one day I stumbled on a recipe showing how easy it is to make baked goat cheese at
home. Now this salad is a staple in my home. I left the base of the salad simple,
but ingredients can easily be added to mix it up.

YIELD: 4 SERVINGS

For the goat cheese:

6 ounces (170 g) goat cheese

⅔ cup (66 g) hazelnut meal

1 teaspoon fresh thyme

¼ teaspoon sea salt

¼ teaspoon freshly ground black pepper

2 tablespoons (30 ml) olive oil

For the dressing:

¼ cup (60 ml) olive oil

¼ cup (60 ml) balsamic vinegar

2 tablespoons (40 g) honey

2 tablespoons (22 g) Dijon mustard

For the salad:

4 to 5 handfuls lettuce

2 cups cherry tomatoes

½ bunch scallions, diced

To make the goat cheese: Preheat the oven to 375°F (190°C, or gas mark 5).

Divide the goat cheese into 8 equal pieces. Lightly pat or roll each piece into a ½-inch (1.2-cm)-thick disk.

Combine the hazelnut meal, thyme, salt, and black pepper in a bowl. One at a time, coat the goat cheese disks with the olive oil. Press each disk into the hazelnut mixture, making sure the goat cheese is fully coated. Place the goat cheese disks in a single layer in a roasting pan and bake until soft and the hazelnuts are roasted, 10 to 12 minutes. Let rest for a couple of minutes before serving.

To make the dressing: In a blender, combine the olive oil, balsamic vinegar, honey, and mustard. Pulse until emulsified.

To make the salad: Toss together the lettuce, cherry tomatoes, and scallions. Toss with the dressing until the vegetables are coated. Divide among 4 salad plates and place 2 warm goat cheese disks on top of each.

HAZELNUT
DARK CHOCOLATE TRUFFLES

I feel as though everyone should have a small repertoire of
recipes that are simple yet always seem to be the most asked about at a party.
These truffles take minimal work and are sure to please a crowd.

YIELD: 24 TO 30 TRUFFLES

12 ounces (340 g) premium dark
  chocolate, chopped
¾ cup (175 ml) heavy cream
1 cup (100 g) hazelnut meal

Place the chocolate in a heat-proof bowl. In a small saucepan, heat the
cream over medium heat until scalded. Pour over the chocolate in a bowl
and let sit for 2 to 3 minutes. Stir the mixture until smooth. If your chocolate still has bits and pieces that didn't melt, place the bowl over a pot of
hot water and use it as a double boiler to melt the remaining chocolate.

Let cool for 1 hour at room temperature. Cover and place in the refrigerator until the chocolate is set enough to scoop and roll, about 2 hours.
If the chocolate has hardened beyond the rolling stage, let it sit on the
counter until it softens again.

Place the hazelnut meal in a small bowl. Using a small cookie scoop or
spoon, scoop 1 to 2 tablespoons (22 g) of chocolate and quickly roll into
a ball in your hands. Toss the truffle in the hazelnut meal, then set on
a baking sheet lined with parchment or waxed paper. Repeat with the
remaining chocolate. Store in an airtight container at room temperature
for 3 to 4 days.

HAZELNUT PUMPKIN MUFFINS

There are certain flavors I always pair with certain nuts:
pecans and sweet potatoes, pistachios and lemon, and hazelnuts and pumpkin.
Fall is my favorite time of year, and these muffins fit right in with the changing leaves
and crisp autumn air. They can be made using either hazelnut meal or flour, but I almost
always make them using the meal to add a bit of crunch to the texture.

YIELD: 12 MUFFINS

1 cup (100 g) hazelnut meal (or flour)

½ cup (50 g) oat flour

½ cup (38 g) teff flour

¼ cup (36 g) sorghum flour

¼ cup (16 g) arrowroot

¼ cup (16 g) cornstarch

1 teaspoon baking soda

1 teaspoon ground cinnamon

½ teaspoon sea salt

¼ teaspoon ground nutmeg

¼ teaspoon ground ginger

1 cup (200 g) pumpkin purée

½ cup (120 ml) maple syrup

¼ cup (56 g) butter, melted and slightly cooled

2 large eggs

Preheat the oven to 375°F (190°C, or gas mark 5). Line a standard 12-cup muffin pan with liners.

In a bowl, stir together the hazelnut meal, oat flour, teff flour, sorghum flour, arrowroot, cornstarch, baking soda, cinnamon, salt, nutmeg, and ginger. In a separate bowl, whisk together the pumpkin purée, maple syrup, melted butter, and eggs. Pour the wet ingredients into the dry ingredients and stir until combined.

Divide the batter among the 12 muffin cups. Bake for 22 to 24 minutes, until the muffins spring back when lightly pressed. Let cool in pan for 5 minutes, then transfer to a wire rack to finish cooling. Store in an airtight container at room temperature for 2 to 3 days, or freeze for extended storage.

WALNUTS

The Midwestern house in which I was raised was well over 100 years old. The old trees that surrounded the house felt slightly majestic, and one in particular, a walnut tree, always kept me intrigued. Once a year, large, slightly leathery green balls fell from the tree. There were two things I knew: They smelled horrendous and they stained my hands an ugly green color. At the time, I had no clue that these green balls were directly related to the walnuts that I loved so much.

There are three basic walnut varieties: English (also known as Persian), white, and black. English walnuts originated in India and are often the variety of walnut found in stores, mainly because of the thin shell that is easily cracked. Black and white walnuts, both native to North America, have a hard shell that is difficult to crack. Black walnuts, while harder to harvest, contain more protein and have a more robust flavor than the English variety. All varieties of walnuts are included in the true nut category in both the botanical and the culinary sense. Walnuts are surrounded by two layers: a hard, wrinkly shell and a leathery green husk. One walnut shell contains two walnuts, separated by a divider.

Many sources suggest removing the thin skin from walnuts to avoid the bitterness that comes from it. However, the thin layer contains some of the nutrients found in walnuts. I have never peeled a walnut before and find the flavor to be just lovely. Though the walnut doesn't have the smooth, buttery taste of the pecan, its flavor is unique and pairs well with slightly earthy flavors.

Walnuts can be found prepackaged, in the bulk bins, or in the coolers at supermarkets and health food stores. It can be a bit harder to find whole walnuts. However, walnut pieces will work just as well as whole walnuts when grinding flour. Walnuts go rancid and stale quickly. Occasionally the walnuts will smell fine but have a soft, stale taste. Always test before grinding or eating.

Walnut Meal

Similar to pecans, walnuts turn to walnut butter rather quickly because of their high level of fats. Walnut meal is best made in small batches in the food processor, pulsing each batch just until it turns into meal. A high-speed blender may work, but I find it is far easier to turn the walnuts into walnut butter that way. Also because of the fats, walnut meal doesn't sift into flour well, and so I use only recipes that work well with meal consistency. Walnut meal is great for use as a crust, pairing with many vegetables, or combining with other flours and meals. Compared with pecans, walnut meal has a slightly earthy and bitter taste that I find unique; it is wonderful in baked goods, especially those with chocolate.

A word of caution: As with sunflower seeds, when walnuts are paired with baking soda in a baked good, a chemical reaction can create a green hue in the finished dish. While this doesn't affect the taste, it can be somewhat unappetizing visually. If the batter is darker (such as with a chocolate dessert), this green hue is less noticeable.

Weights and Measurements

1 cup walnuts = 100 g

1 cup walnut meal = 100 g

1 cup (100 g) walnuts = 1 cup (100 g) walnut meal

WALNUT COCOA BROWNIES

There are hundreds, if not thousands, of brownie recipes. On top of that,
there are two brownie camps: those who love dense, fudgy brownies, and those
who love thick, cakey brownies. For quite a while, I always chose cakey brownies, until
I tried a dense version that my mother made. Those brownies were perfectly sweetened,
not too rich, and fudgy. This brownie recipe is an adaptation of her recipe,
using walnuts to add a bit of crunch to the texture.

YIELD: 12 BROWNIES

½ cup (50 g) walnut meal

¼ cup (32 g) arrowroot

¾ cup (150 g) organic cane sugar

6 tablespoons (30 g) unsweetened
  cocoa powder

¼ teaspoon sea salt

¼ cup (60 ml) walnut oil

1 teaspoon vanilla extract

1 large egg

Preheat the oven to 300°F (150°C, or gas mark 2). Lightly grease an
8 × 8-inch (20 × 20-cm) baking pan.

In a bowl, stir together the walnut meal, arrowroot, cane sugar, cocoa
powder, and salt. In a separate bowl, whisk together the walnut oil,
vanilla, and egg. Pour the wet ingredients into the dry ingredients and
stir until well combined.

Spread the brownie batter into the prepared baking pan. Bake for 45 to
50 minutes, until the brownies are pulling away from the sides of the
pan and are set in the middle. Let cool completely before slicing into
12 squares. Store in an airtight container at room temperature for
2 to 3 days.

TWICE-BAKED SWEET POTATOES
WITH WALNUT CRUMBLE

By now you may have noticed my penchant for sweet potatoes.
During the winter, especially while residing in the Midwest, I lived on these nutritious
tubers. Most of the time, sweet potatoes will play a key role in the main dish, but they are
also wonderful as a side. These twice-baked sweet potatoes are a tasty side that comes
together easily and can be multiplied depending on how many you are feeding.

YIELD: 2 SIDE-DISH SERVINGS

1 medium sweet potato

1 ounce (28 g) blue cheese

1 tablespoon (15 ml) heavy cream

½ clove garlic, minced

¼ cup (25 g) walnut meal

¼ teaspoon sea salt

¼ teaspoon freshly ground black pepper

1 teaspoon olive oil

Preheat the oven to 400°F (200°C, or gas mark 6).

Pierce the sweet potato 4 or 5 times with a fork. Place on a baking sheet and bake until soft, 40 to 45 minutes. Let cool slightly.

Cut the sweet potato in half and scoop out the flesh into a medium bowl, leaving a ¼-inch (3-mm) shell of sweet potato around the edges of the skin. Add the blue cheese, heavy cream, and garlic to the sweet potato in the bowl. Whip the mixture until well combined and spoon it into the sweet potato shells.

In a bowl, combine the walnut meal, salt, and pepper. Rub in the olive oil until the walnuts are coated. Sprinkle the mixture over the sweet potatoes. Bake for 20 to 25 minutes, until the walnuts are browning and the cheese is melted.

SUN-DRIED TOMATO AND WALNUT PÂTÉ

I'm always on the hunt for great appetizers that don't involve
meat or cheese. This simple spread is perfect to serve with crackers or toasted
bread and comes together quickly. I always use dry-packed sun-dried tomatoes
to avoid extra oil. You can always use oil-packed sun-dried tomatoes,
but the oil will result in a slightly different texture.

YIELD: 1 CUP (433 G)

1 cup (108 g) sun-dried tomatoes
   (dry-packed)

1 cup (235 ml) water

2 cloves garlic, peeled

½ cup (50 g) walnut meal

2 teaspoons fresh rosemary

1 tablespoon (15 ml) fresh lemon juice

2 teaspoons honey

1 teaspoon freshly ground black pepper

½ teaspoon sea salt

In a small bowl, combine the sun-dried tomatoes with the water and let
sit until the tomatoes are tender, about 2 hours. Drain and reserve the
excess water. Pulse the garlic cloves in a food processor until minced.
Add the drained sun-dried tomatoes, walnut meal, rosemary, lemon juice,
honey, pepper, and salt. Pulse until the mixture comes together. Add the
sun-dried tomato water to thin the texture, if desired.

This dip can be made up to 1 day ahead. Simply store in an airtight container in
the refrigerator until ready to use.

PECANS

I find it hard to play favorites among grains, as I enjoy them all, but nut varieties are a different story. I love pecans, and I'm not very bashful about it. My favorite granola has only five ingredients, one of which is pecans. Most recipes I make that incorporate sweet potatoes or butternut squash are usually accompanied by pecans, and my favorite snack is a handful of pecans and dark chocolate chips. I can't get enough of this nut's sweet, smooth taste.

Pecans, a species of hickory, come from a deciduous tree that is native to parts of the United States and Central America. These nuts are actually a stone fruit, and the part we know as the nut is the seed for the tree. Pecans have a hard exterior, making them a tough nut to crack. This characteristic is also behind how pecans got their name. *Pecan* is of Algonquin origin, meaning "all nuts requiring a stone to crack."

The United States is one of the top growers of pecans; however, the nuts are grown around the world. Pecans can be purchased both with and without the shell in grocery stores, health food stores, and online. Pecans are also sold in halves and pieces, and it does not matter which you purchase for making meal. To ensure freshness, look for pecans that have been refrigerated.

Pecan Meal

Compared with almonds, pecans are harder to grind into a fine flour because they have one of the highest amounts of fats. When grinding, those fats get released into an oil that very easily turns pecans into pecan butter. Pecans are best ground in a food processor in small batches, only pulsed a few times. Nut grinders also work and can help prevent turning the pecans into pecan butter.

Pecan meal is one of the sweeter nut meals, making it a great addition to baked goods, especially cookies and quick breads. The meal can also be used for creating a slightly crunchy and nutty topping for casseroles and even macaroni and cheese.

Check the pecans before grinding. I find pecans go stale quicker than other nuts, and the texture and taste of stale nuts is not very pleasing. Pecans and pecan meal should always be stored in the refrigerator or freezer.

Weights and Measurements

1 cup pecans = 100 g

1 cup pecan meal = 100 g

1 cup (100 g) pecans = 1 cup (100 g) pecan meal

CAULIFLOWER GRATIN WITH PECAN MEAL CRUST

During holiday meals, I often skip over the main dish
and load my plate up with side dishes. I'm always drawn to the dishes
packed with vegetables, and it never hurts when they happen to be covered
with cheese. The pecan crust in this gratin adds a bit of crunch along with flavor
that plays well with the roasted cauliflower.

YIELD: 4 TO 6 SERVINGS

For the cauliflower:

5 cups (420 g) small cauliflower florets
  (1 medium head)

2 tablespoons (30 ml) olive oil

½ teaspoon sea salt

For the sauce:

1½ tablespoons (21 g) butter

½ cup (80 g) minced onion

1 tablespoon (8 g) cornstarch

2 teaspoons minced fresh sage

¾ cup (175 ml) 2% or whole milk

¾ cup (74 g) shredded mozzarella

¼ cup (69 g) ricotta

For the topping:

1 tablespoon (14 g) butter, melted

¾ cup (75 g) pecan meal

½ cup (8 g) grated Parmesan

To make the cauliflower: Preheat the oven to 375°F (190°C, or gas mark 5).

In a bowl, toss together the cauliflower, olive oil, and salt. Spread in a single layer on a baking sheet and roast until tender, 20 to 25 minutes. Meanwhile, make the sauce.

To make the sauce: Melt the butter in a saucepan over medium-low heat. Stir in the onion and cook until translucent, 5 to 6 minutes. Stir in the cornstarch and sage, and heat for 1 more minute. Add the milk and cook until the sauce thickens, stirring frequently. Once thick, remove from the heat and stir in the mozzarella and ricotta.

In a 2½-quart (2.4-L) baking dish, combine the cheese sauce and cauliflower, stirring until the cauliflower is covered.

To make the topping: In a medium bowl, combine the butter with the pecan meal and Parmesan. Sprinkle over the cauliflower.

Bake for 22 to 25 minutes, until the cheese sauce is bubbling and the pecan meal is slightly crispy.

Broccoli or Brussels sprouts are also lovely with this dish instead of cauliflower—simply adjust the roasting time until tender.

PECAN-CRUSTED GREEN TOMATO SANDWICHES

During the early summer months, I anxiously watch the
tomatoes grow, waiting for them to be juicy and ripe. Occasionally, however,
I cannot wait and pluck a green one off the vine. This sandwich is my take
on the traditional fried green tomato, with the help of pecan meal for
extra flavor and a zesty chipotle ranch dressing.

YIELD: 4 SANDWICHES

For the dressing:

½ cup (112 g) whole-milk plain Greek yogurt

2 tablespoons (6 g) minced fresh chives

1 tablespoon (4 g) minced fresh parsley

1½ teaspoons minced fresh dill

½ small clove garlic, minced

½ teaspoon freshly ground black pepper

½ teaspoon sea salt

2 tablespoons (30 ml) buttermilk

½ teaspoon ground chipotle chili powder, plus more as desired

For the sandwiches:

1 large (260 g) green heirloom tomato

2 tablespoons (30 ml) heavy cream

1 large egg yolk

⅔ cup (66 g) pecan meal

½ teaspoon sea salt

½ teaspoon freshly ground black pepper

2 tablespoons (30 ml) olive oil

Lettuce leaves

4 hamburger buns

To make the dressing: Whisk together the yogurt, chives, parsley, dill, garlic, pepper, salt, and buttermilk. Add the ½ teaspoon chipotle powder and taste. Add more chipotle powder as needed to reach the desired flavor and heat. Place in the refrigerator until ready to use.

To make the sandwiches: Slice the tomato into four ½-inch (1.2-cm)-thick slices. Set up a breading station by whisking together the cream and egg yolk in one shallow bowl, and combining the pecan meal, salt, and pepper in another shallow bowl. Dunk the tomato slices into the cream mixture and then transfer to the pecan mixture. Press the slices down to ensure the pecan meal sticks.

Heat the olive oil in a large skillet over medium-low heat. Add the pecan-crusted tomatoes to the skillet and cook on each side until the nuts are browning and the tomatoes are tender.

Layer lettuce leaves and 1 slice of heirloom tomato on the bottom halves of each bun. Spread 1 to 2 tablespoons of dressing on the top half of each bun and place over the tomatoes. Serve warm.

Because heirloom tomatoes differ greatly in size, you may need 1 to 2 tablespoons (15 to 30 ml) more heavy cream mixture and 2 to 4 tablespoons (12.5 to 25 g) more pecan mixture.

PECAN MEAL BANANA BREAD

As much as I love bananas that are barely ripe, I am the biggest
culprit when it comes to letting them overripen. It seems that every time I turn around,
I have bananas that are perfect for banana bread. My favorite addition to banana bread
has always been pecans, and so making banana bread partially out of pecan meal
creates a wonderful flavor and gives the bread a slight crunch. Walnut meal
also works beautifully in this recipe.

YIELD: ONE 8-INCH (20-CM) LOAF

1¼ cups (125 g) pecan meal

¼ cup (32 g) arrowroot

¼ cup (32 g) tapioca starch

¼ cup (36 g) brown rice flour

½ cup (68 g) sorghum flour

¼ cup (38 g) teff flour

1 teaspoon baking soda

½ teaspoon sea salt

2 teaspoons ground cinnamon

1 teaspoon ground nutmeg

½ teaspoon ground ginger

1¼ cups (240 g) banana purée (from
    2 or 3 bananas)

½ cup (120 ml) maple syrup

¼ cup plus 2 tablespoons (90 ml) pecan or
    walnut oil

2 large eggs

Preheat the oven to 375°F (190°C, or gas mark 5). Lightly grease a
5 × 8-inch (13 × 20-cm) bread pan.

In a large bowl, stir together the pecan meal, starches, flours, baking
soda, salt, cinnamon, nutmeg, and ginger. In a separate bowl, whisk
together the banana purée, maple syrup, oil, and eggs. Pour the wet
ingredients into the dry ingredients and stir until just combined.

Pour the batter into the prepared bread ban and bake for 40 minutes,
until dark brown and a knife inserted in the center comes out clean.
Let cool for 10 minutes in the pan, then transfer to a wire rack to
finish cooling. Store in an airtight container at room temperature for
2 to 3 days, or freeze for extended storage.

Resources

Many items listed in this book would have been impossible to find at local stores while I was growing up in rural Illinois. Stores that sell online are a great way to find these items at reasonable prices.

Bob's Red Mill

www.bobsredmill.com
(800) 349-2173

In stores and online, Bob's Red Mill is a great one-stop shop for nearly all the items listed in this book. They sell organic, gluten-free, and conventional grains, seeds, nuts, and legumes. Many of their products are sold in health food stores and larger supermarkets.

Arrowhead Mills

arrowheadmills.elsstore.com
(800) 434-4246

Arrowhead Mills offers a small selection of organic grains, legumes, and seeds in health food stores, larger supermarkets, and online.

Nuts Online

www.nuts.com
(800) 558-6887

If you're looking to order a mix of nuts, legumes, and grains in smaller quantities, Nuts Online is a good place to start. Some of the harder-to-find grains, such as teff and sorghum, can be purchased by the pound. However, not all the products are organic.

Jovial Foods (Einkorn)

store.jovialfoods.com
(877) 642-0644

Einkorn berries are still fairly new to the scene and not many stores, local or online, sell the berries. Jovial Foods sells wheat berries in small quantities.

Lundberg Family Farms

www.lundberg.com
(530) 538-3500

As a fairly new California citizen, I've become aware of all the great companies to buy from locally. Lundberg Farms is a great place to order organic long-grain, short-grain, and sweet brown rice. Its rice can be found in health food stores and supermarkets around the country and can be ordered from its site in small quantities and in bulk.

Acknowledgments

Of all the things I expected to do in my life, writing a book wasn't always on the list. And yet, here I am, and I have many to give big hugs to and thank.

To my new husband, Mike, who holds me tight when I need comfort and who acts as my biggest cheerleader for all my ideas. I can't imagine having a better person on my side through this crazy journey.

To my parents and grandparents, for nurturing my extensive curiosity and giving me more opportunities than a girl could ever hope for. To my father, whose honesty keeps me growing in photography, and to my mother, for talking all things food and testing recipes for the book.

To Sarah Kieffer, Nicole Gulotta, Erika and Stephanie Cline, Tim and Shanna Mallon, and April Holland, for helping me test and work with my somewhat interesting recipes.

To Jill Alexander from Fair Winds Press, for reaching out with this fun idea for a cookbook and being so wonderful to work with on everything.

And of course, to my readers: Because of you, I've continued to cook and share recipes, without which this book would not have been possible.

About the Author

Photographer and Web designer Erin Alderson is the voice behind Naturally Ella, a whole foods, vegetarian blog that features accessible, healthy recipes. Erin's work has been featured on The Kitchn, *Food and Wine*, Food52, and *Bon Appétit*. When not creating a mess in the kitchen, Erin can be found in the mountains hiking or snowboarding. She currently resides in Sacramento, California, with her husband, Mike, and her husky, Radar.

Index